Hierarchical Bayesian

Studies in Fuzziness and Soft Computing, Volume 170

Editor-in-chief
Prof. Janusz Kacprzyk
Systems Research Institute
Polish Academy of Sciences
ul. Newelska 6
01-447 Warsaw
Poland
E-mail: kacprzyk@ibspan.waw.pl

Martin Pelikan

Hierarchical Bayesian Optimization Algorithm

Toward a New Generation
of Evolutionary Algorithms

 Springer

Martin Pelikan
University of Missouri
Dept. of Mathematics
and Computer Science
St. Louis, MO 63121
USA
E-mail: pelikan@cs.umsl.edu

ISSN print edition: 1434-9922
ISSN electronic edition: 1860-0808
ISBN 978-3-642-06273-5   e-ISBN 978-3-540-32373-0

Springer is a part of Springer Science+Business Media
springeronline.com

© Springer-Verlag Berlin Heidelberg 2010
Printed in Germany

Cover design: E. Kirchner, Springer Heidelberg
Printed on acid-free paper   62/3141/jl- 5 4 3 2 1 0

To my parents, Jaroslava and Peter

Foreword

At the 1997 Genetic Programming Conference at Stanford University, a self-described "long red-haired guy" walked up to me and introduced himself. His name was Martin Pelikan, he came from Slovakia, and he wanted to visit my lab at Illinois. He was doing interesting work, and he seemed like a bright enough fellow, so we exchanged e-mail addresses and worked on overcoming the logistical problems of a visit. Doing so took longer than I expected, but on September 2, 1998 he arrived in Champaign and immediately got to work. The initial visit turned into a stint as a PhD student, the stint turned into a dissertation, and that dissertation has now been transformed into the book before you. It is a remarkable piece of work.

Simply stated Martin has shown how to take (a) population-based search, (b) Bayesian networks, (c) niching or clustering techniques and (d) compact representations and create a fairly general purpose optimizer of remarkable breadth and capability. His system is called the hierarchical Bayesian optimization algorithm (hBOA). I have used the term competent to characterize genetic and other search procedures that solve a large class of hard problems quickly, reliably, and accurately, and the book before you is a comprehensive description of the thinking that went into the design of a competent hBOA, its precursors, the analytical theory of hBOA operation, and empirical tests of hBOA on both bounding test functions and problems of interest in the real world; it is a thorough discussion of what in my view is the most competent solver to have emerged from these and related lines of thinking.

The book is original in many respects, so it is difficult to single out any one of its many contributions. Four items do deserve particular comment:

- Decomposition of the problem of hierarchical competence.
- Analysis of hBOA scalability.
- Characterization of a bounding class of hierarchically difficult problems.
- Demonstration of hBOA in well-chosen test and real-world functions.

The key that opened the door to a fuller understanding of the everyday solution of hierarchically difficult problems was a tripartite decomposition of their solution. Martin's work clearly articulates the need for (a) capable learning of important substructures, (b) preservation of alternate competing substructures that appear to solve a problem, and (c) their compact representation. This decomposition is the subject of a recent patent filing, but more importantly, doing all three things simultaneously was absolutely essential to the progress made.

This qualitative decomposition of function was critical, but this study was no mere exercise in clever intuition. Martin followed his design creativity with an especially careful analysis of algorithmic scalability by better understanding the population size and run duration necessary to achieve high quality solutions.

These bounding results would not have been as meaningful had Martin not carefully considered boundedly difficult test functions for his initial testing. Parsimonious testing at the edge of the design envelope ensures that hBOA performs as well if not better on problems easier than those tested. Real-world problems sometimes offer up their own surprises, but extensive testing on spin glasses and solvable instances of maximum satisfiability problems offered more than a little assurance that the design methodology adopted is sound.

These extraordinary results would be sufficient for me to recommend this book to genetic algorithmists and artificial evolutionaries around the world, but the melding of population thinking, traditional topics of artificial intelligence and machine learning, and analytics drawn from population biology (both real and artificial) suggest that this text may be of interest to a much broader audience than the usual treatise on genetic algorithms. The communities mentioned above have a long history of talking past one another, but Martin's work hurries past the intellectual chauvinism to integrate important contributions of many fields into a helluva solver. I urge anyone interested in the design, theory, implementation, and application of a broadly competent solver to buy this book, read it, and engage its lessons in thought or in practice.

Urbana, Illinois, USA   *David E. Goldberg*
January 2005

Preface

A black-box optimization problem may be defined by specifying (1) a set of all potential solutions to the problem and (2) a measure for evaluating the quality of each candidate solution. The task is to find a solution or a set of solutions that perform best with respect to the specified measure. An important feature of black-box optimization is that a black-box optimizer does not know anything about the semantics of potential solutions or the relationship between potential solutions and the evaluation procedure. Potential solutions can represent anything from aircraft wings, to chess strategies, to musical compositions, to job schedules, or to configurations of a molecule. The evaluation procedure can be based on an experiment in a wind tunnel, a traffic simulation, or a computer procedure. The only way how a black-box optimizer can learn something about the problem is to sample new solutions, evaluate them, and process the results of the evaluation. Since many challenging real-world problems can be formulated as black-box optimization problems, the design of automated, robust, and scalable black-box optimization methods is one of the most important challenges in computational optimization.

The primary goal of this book is to design an advanced black-box optimization algorithm for automated, robust, and scalable solution of a broad class of real-world problems without the need for interaction with the user or problem-specific knowledge in addition to the set of potential solutions and the performance measure. Despite that the book tries to alleviate the need for expensive manual problem analysis, the methods presented here enable the use of prior problem-specific knowledge of various forms. Consequently, if the practitioner has additional knowledge about the problem, this can likely be used for further efficiency enhancement.

To meet this difficult challenge, the book derives inspiration from the way the best problem solvers – humans – solve their problems. To simplify problems and to develop tractable models of complex systems, humans often break up the problem or system into several subproblems or subsystems. This can be done on a single or multiple levels, but the basic idea remains the same – decompose the big problem into smaller subproblems, solve the subproblems (either directly or via yet another level of decomposition), and combine the results to form the solution to the entire problem. Although the basic idea of using decomposition in problem solving is simple, decomposition can be used across an enormously diverse spectrum of areas from mathematical theorem proving to engineering design, to computation of magnetic properties of complex physical systems, to development of models of complex ecological systems, and to music composition.

The algorithms developed in this book borrow ideas from genetic and evolutionary computation, machine learning, and statistics. The work is most closely related to genetic and evolutionary computation, which provides us with the concepts of population-based search, directed exploration by combining features of promising solutions, diversity preservation techniques, and facetwise theory and design. New operators are designed for exploration of the space of all potential solutions based on methods adopted from machine learning and statistics to ensure automatic discovery and exploitation of single-level and hierarchical decomposition in genetic and evolutionary algorithms. The combination of genetic and evolutionary computation with the methods of machine learning and statistics enables quick, accurate, and reliable solution of a large class of nearly decomposable and hierarchical problems, many of which are intractable by other common optimization techniques.

This book is intended for a wide audience and it does not require the reader to be an expert in genetic and evolutionary computation, machine learning, or statistics. Basics in mathematics and statistics should provide sufficient background to understand most material in this book. Basic knowledge of genetic and evolutionary computation and machine learning is not necessary, but it can allow the reader to proceed faster.

Road Map

The book is divided into eight chapters. The first two chapters introduce genetic algorithms (GAs), provide motivation for the design of competent GAs, and introduce the basic concepts of probabilistic model-building genetic algorithms (PMBGAs). Chapters 3 and 4 present and analyze the Bayesian optimization algorithm (BOA), which is theoretically and empirically shown to solve decomposable problems of bounded difficulty in a scalable manner. Chapters 5 and 6 motivate the use of hierarchical decomposition for complexity reduction, propose a class of challenging problems for hierarchical optimizers, and extend the original BOA to solve difficult hierarchical problems. The book closes by presenting experimental results on two classes of real-world problems and providing the conclusions. The following subsections present the content of each chapter in greater detail.

Chapter 1: From Genetic Variation to Probabilistic Modeling

Chapter 1 describes the basic GA procedure and argues that in the design of competent black-box optimization techniques based on problem decomposition, there are several important lessons to learn from GA research. The chapter presents a simple GA simulation by hand to build intuition about the dynamics of GAs. Additionally, the simulation provides motivation for the use of probabilistic models in guiding exploration more generally.

Chapter 2: Probabilistic Model-Building Genetic Algorithms

Chapter 2 surveys PMBGAs, which guide exploration in black-box optimization by building a probabilistic model of promising solutions found so far and sampling the built model to generate new candidate solutions. Basic definitions are provided to present PMBGAs within a unified framework. The chapter focuses on methods working in a discrete domain; other representations are discussed briefly.

Chapter 3: Bayesian Optimization Algorithm

Chapter 3 proposes BOA, which uses Bayesian networks to decompose the problem and sample new candidate solutions. Next, the chapter describes basic methods for learning and sampling multiply connected Bayesian networks. The chapter finishes by presenting initial experimental results indicating good scalability of BOA on decomposable problems of bounded difficulty.

Chapter 4: Scalability Analysis

To support the promising empirical results presented in the previous chapter, Chapter 4 develops a facet-wise scalability theory for BOA. As a measure of computational complexity of BOA, the chapter considers the total number of candidate solutions that must be evaluated until reliable convergence to an optimum. The total number of evaluations is computed by estimating an upper bound on the sufficient population size and an expected number of generations until convergence. It turns out that if a problem can be decomposed into subproblems of bounded order, the number of evaluations until BOA converges to the optimum with high confidence grows subquadratically or quadratically with problem size. The theory is verified by a number of experiments.

Chapter 5: The Challenge of Hierarchical Difficulty

Chapter 5 discusses the use of hierarchical decomposition as opposed to decomposition on only a single level and argues that hierarchical decomposition is a powerful tool for reducing problem complexity. The chapter identifies important features that must be incorporated into BOA to allow the algorithm to exploit hierarchical decomposition effectively: (1) proper decomposition, (2) chunking, and (3) preservation of alternative solutions.

Next, the chapter proposes a class of challenging hierarchical problems called hierarchical traps. Hierarchical traps bound the class of problems solvable by optimization algorithms capable of automatic discovery and exploitation of hierarchical decomposition and can be used as the test of fire for those algorithms. Hierarchical traps are extremely difficult if they are approached on a single level or with local operators. However, with an effective use of hierarchical decomposition, hierarchical traps can be solved in low-order polynomial time after evaluating only a subquadratic number of candidate solutions.

Chapter 6: Hierarchical Bayesian Optimization Algorithm

Chapter 6 describes how to incorporate the important features of a successful hierarchical optimizer into BOA. Decomposition and chunking are incorporated by using local structures that ensure compact representation of Bayesian networks for problems with high-order and complex interactions. Preservation of alternative solutions is guaranteed by using a procedure called restricted tournament replacement to update the population of candidate solutions using newly generated solutions. Bayesian networks with local structures and restricted tournament replacement form the basis of the hierarchical Bayesian optimization algorithm (hBOA).

hBOA is then shown to solve hierarchical traps and other hierarchical problems in a scalable manner. To provide a bound on the number of evaluations required for solving hierarchical problems, BOA scalability theory is extended to account for optimization over multiple levels of decomposition. The extended theory is verified with empirical results.

Chapter 7: Hierarchical BOA in the Real World

Chapter 7 applies hBOA to two important classes of real-world problems: (1) two- and three-dimensional Ising spin glasses, and (2) maximum satisfiability (MAXSAT). The results indicate that hBOA provides competitive or better results than specialized techniques designed to solve specifically these classes of problems and that it can solve a number of difficult problem instances that cannot be solved by the specialized techniques.

Chapter 8: Summary and Conclusions

Chapter 8 summarizes main contributions of this work and provides important conclusions.

Acknowledgments

Foremost, I would like to thank my parents and the rest of my closest family, my brother Juraj, my sister Daniela, and my godparents Dušanko and Elenka. I am grateful for having the family I have, it's perfect. I would also like to thank Angelica L. Farnum whom I consider as a part of my family, for standing beside me along the way.

I would certainly not complete this book without David E. Goldberg whom I value as a great teacher, source of inspiration, coworker and friend. Those who know Dave Goldberg and his work will quickly recognize his strong influence on the research presented here. Most of this work was completed at Dave Goldberg's Illinois Genetic Algorithms Laboratory (IlliGAL) at the University of Illinois at Urbana-Champaign. Working at IlliGAL was a great professional experience and a lot of fun, too. I would like to thank all members and visitors of the lab whom I could work with. I am particularly thankful to Jacob L. Borgerson, Martin V. Butz, Erick Cantú-Paz, Fernando G. Lobo, Franz Rothlauf, Kumara Sastry, Shigeyoshi Tsutsui, and Clarissa Van Hoyweghen.

A number of researchers contributed to this book by providing valuable comments, answering my questions, and sharing their views. To name a few, I acknowledge Fabien Alet, Peter A. N. Bosman, David Chickering, Kalyanmoy Deb, Georges Harik, David Heckerman, Hillol Kargupta, Pedro Larrañaga, Thilo Mahnig, Heinz Mühlenbein, Bart Naudts, Jiri Ocenasek, Josef Schwarz, Dirk Thierens, Simon Trebst, Matthias Troyer, and Richard A. Watson.

I would also like to thank my colleagues at the Swiss Federal Institute of Technology (ETH) in Zürich and the University of Missouri at St. Louis, where I finished writing this book.

The work was sponsored by the Air Force Office of Scientific Research, Air Force Materiel Command, USAF, under grants F49620-97-1-0050 and F49620-00-0163. Research funding for this project was also provided by a grant from the U.S. Army Research Laboratory under the Federated Laboratory Program, Cooperative Agreement DAAL01-96-2-0003, and a grant from the National Science Foundation under grant DMI-9908252. Some experiments were done using the computational facilities of the National Center for Supercomputing Applications (NCSA) at the University of Illinois at Urbana-Champaign. Some experiments were also done using the Asgard cluster at the Swiss Federal Institute of Technology in Zürich (ETHZ). Part of this work was also sponsored by the Research Board at the University of Missouri.

The US Government is authorized to reproduce and distribute reprints for Government purposes notwithstanding any copyright notation thereon. The views and conclusions contained herein are those of the author and should not be interpreted as necessarily representing the official policies or endorsements, either expressed or implied, of the Air Force Office of Scientific Research, the National Science Foundation, or the U.S. Government.

St. Louis, Missouri, USA
January 2005   *Martin Pelikan*

Contents

1

From Genetic Variation
to Probabilistic Modeling

Genetic algorithms (GAs) [53, 83] are stochastic optimization methods inspired by natural evolution and genetics. Over the last few decades, GAs have
been successfully applied to many problems of business, engineering, and science [56]. Because of their operational simplicity and wide applicability, GAs
are now becoming an increasingly important area of computational optimization.

There are several GA concepts that will be used in the design of a competent black-box optimization method capable of exploiting problem decomposition: population-based search, exploration by combining bits and pieces
of promising solutions, and facetwise GA theory. The purpose of this chapter is to review the basic GA procedure and motivate the use of methods
for learning and sampling probabilistic models instead of traditional variation
operators inspired by genetics.

The chapter starts by discussing the general form of a black-box optimization problem. Section 1.2 describes the basic GA procedure. Section 1.3
presents a sample GA run. The simulation motivates the use of probabilistic
recombination, which consists of building a probabilistic model of promising
solutions and sampling the built model to generate new candidate solutions.
Section 1.4 presents one approach to probabilistic recombination and relates
the presented approach to the traditional one. Finally, Sect. 1.5 discusses limitations of the presented probabilistic variation operator and indicates how
this can be extended to cope with more complex problems.

1.1 Black-Box Optimization

An optimization problem may be defined by specifying (1) a set of all potential
solutions to the problem and (2) a measure to evaluate the performance of
each candidate solution with respect to the objective. The goal is to find a
solution or a set of solutions that perform best with respect to the specified
performance measure. For example, in maximum satisfiability (MAXSAT),

Martin Pelikan: *Hierarchical Bayesian Optimization Algorithm*, StudFuzz **170**, 1–12 (2005)
www.springerlink.com

each candidate solution represents an interpretation of all propositions (a list of truth values of all variables or propositions) and the quality of a solution can be defined as the number of satisfied clauses using the interpretation encoded by the solution. In the design of an aircraft wing, a solution can be represented by a set of parameters that specify the shape of the wing and the performance of each parameter set can be determined by an experiment in a wind tunnel. In the design of an algorithm for playing chess, a solution can be represented by a set of condition-action rules, and the performance of each such set can be defined as the portion of games won against other candidate strategies.

In black-box optimization, there is no information about the relation between the semantics of solutions and the performance measure. The only way of learning something about this relation is to *sample new candidate solutions and evaluate them.* The task of finding the best solution to a black-box optimization problem is extremely difficult. To illustrate the difficulty of black-box optimization, imagine you are asked to implement a program in an unknown programming language given only the syntax of the language and a procedure that evaluates how good each valid program is.

Black-box optimization techniques proceed in iterations where in each iteration, new candidate solutions are first generated, and the results of the evaluation of these solutions are then used to update the sampling procedure. Ideally, the quality of sampled solutions should improve over time, yielding a global optimum after a reasonable number of iterations. A hill climber, for instance, starts in a randomly generated solution. In each iteration, it explores the neighborhood of the current best candidate solution by perturbing the solution in some way. If the perturbations reveal a candidate solution that is better than the current one, the better solution serves as the starting point for further exploration.

The way in which an optimization method samples new candidate solutions and exploits the result of the evaluation of these new solutions limits the class of problems that the method can solve efficiently. For instance, using local operators in the hill climber limits the applicability of the algorithm to problems that contain only a few basins of attraction, or problems where an approximate location of the global optimum is known in advance. The following section introduces GAs as one of the approaches to black-box optimization, where the search for an optimum is driven by ideas inspired by the Darwinian survival of the fittest and the Mendelian inheritance of parental traits.

1.2 Genetic Algorithms

Genetic algorithms (GAs) [53, 83, 108] approach black-box optimization by evolving a population of candidate solutions using operators inspired by natural evolution and genetics. Maintaining a *population of solutions* – as opposed to a *single solution* – has several advantages. Using a population allows a simultaneous exploration of multiple basins of attraction, it allows for statistical decision making based on the entire sample of promising solutions even when the evaluation procedure is affected by external noise, and it enables the use of learning techniques to identify problem regularities.

GAs pose no significant prior restrictions on the representation of candidate solutions or the performance measure. Representations can vary from binary strings to vectors of real numbers, to permutations, to production rules, to schedules, and to program codes. Performance measures can be based on a computer procedure, a simulation, an interaction with a human, or a combination of the above. Performance can also be evaluated using a partial order operator on the space of all candidate solutions, which compares relative quality of two solutions. Additionally, evaluation can be probabilistic and it can be affected by external noise. Nonetheless, for the sake of simplicity, the rest of this chapter assumes that solutions are represented by binary strings of fixed length and that the performance of each candidate solution is represented by a real number called *fitness*. The task is to find a binary string or a set of binary strings with the highest fitness.

The remainder of this section describes the basic GA procedure and some terms necessary for understanding the rest of this chapter. The basic GA procedure consists of the following key ingredients:

- **Initialization.** GAs usually generate the first population of candidate solutions randomly according to a uniform distribution over all possible solutions.
- **Selection.** Each iteration starts by selecting a set of promising solutions from the current population based on the performance of each solution. Various selection operators can be used, but the basic idea of all selection methods is the same – *make more copies of solutions that perform better at the expense of solutions that perform worse.* Two selection methods will be used in this book: tournament selection and truncation selection. Tournament selection selects one solution at a time by first choosing a random subset of candidate solutions from the current population and then selecting the best solution out of this subset. Random tournaments are repeated until there are sufficiently many solutions in the selected population. The size of the tournaments determines selection pressure – the larger the tournaments, the higher the pressure on the quality of each solution. Truncation selection selects the best $1/s$ of the current population where s determines selection pressure. To select a population of the same size as the population before selection, s copies of the selected solutions can be created.
- **Variation.** Once the set of promising solutions has been selected, new candidate solutions are created by applying recombination (crossover) and mutation to the promising solutions. Recombination combines subsets of promising solutions by exchanging some of their parts; mutation perturbs the recombined solutions slightly to explore their immediate neighborhood.

(a) One-point crossover.   (b) Uniform crossover.

Fig. 1.1. An illustrative example of two common two-parent crossover operators. In one-point crossover, the tails are exchanged after a randomly chosen position. In uniform crossover, the bits in each position are exchanged with probability 50%

Most of the commonly used crossover operators combine partial solutions between pairs of promising solutions with a specified probability. For example, one-point crossover first randomly selects a single position in the two strings and exchanges the bits on all the subsequent positions (see Fig. 1.1a). On the other hand, uniform crossover exchanges the bits in each position with probability 50% (see Fig. 1.1b). Some recombination operators combine all candidate solutions after selection. For example, population-wise uniform crossover creates new candidate solutions by shuffling the bits in each position among the selected set of promising solutions. For binary strings, bit-flip mutation is usually used. Bit-flip mutation proceeds by flipping each bit with a fixed probability. The probability of flipping each bit is quite small, so only small changes are expected to occur. Recombination is the primary source of variation in most GAs; these GAs are often referred to as *selectorecombinative GAs*.

- **Replacement.** After applying crossover and mutation to the set of promising solutions, the population of new candidate solutions replaces the original one and the next iteration is executed (starting with selection), unless termination criteria are met. For example, the run can be terminated when the population converges to a singleton, the population contains a good enough solution, or an upper bound on the number of iterations has been reached.

In explaining and discussing the dynamics and limitations of GAs, several additional terms will be used. A *partial solution* denotes specific bits on a subset of string positions. For example, if we consider 100-bit binary strings, a 1 in the second position and a 0 in the seventh position is a partial solution. Competitors of a partial solution are all partial solutions that are specified in the same positions as the partial solution but that differ from the partial solution in at least one of the specified bits. A *partition* denotes a subset of string positions. Finally, the partition corresponding to a particular partial solution is defined as the subset of string positions that are specified in this partial solution.

Common GA terminology is listed in Table 1.1. In the rest of the book, we will use terminology consistent with that listed in the above table.

Table 1.1. Common GA terminology

Term	Alternative Name(s)
candidate solution	individual, chromosome, string
decision variable	variable, locus
value of decision variable	bit, allele
performance function	fitness function
performance value	fitness, fitness value
population of promising solutions, $S(t)$	parent population, parents, selected solutions
population of new solutions, $O(t)$	offspring population, offspring
iteration	generation

1.3 Simulation: Onemax and Population-Wise Uniform Crossover

By applying selection and crossover to the population of candidate solutions, GAs can process a large number of partial solutions in parallel [53, 83]. The processing of partial solutions consists of (1) deciding between competing partial solutions in the same partition, (2) making more copies of those partial solutions that perform better than their competitors, and (3) combining partial solutions to ensure effective exploration of the search space. Selection ensures that superior partial solutions will be given more copies in the selected population. Crossover ensures that partial solutions propagated by selection are combined to explore new regions in the search space. The massively parallel processing of partial solutions – often referred to as *implicit parallelism* [83] – is one of the most important strengths of GAs.

This section presents a simple GA simulation by hand. The purpose of presenting the simulation is twofold. First, the simulation attempts to build intuition about the processing of partial solutions in GAs. Second, the simulation motivates the use of probabilistic models to guide the sampling of new candidate solutions in GAs.

Onemax is defined as the sum of the bits in the input binary string:

$$f_{onemax}(X) = \sum_{i=1}^{n} X_i \, , \tag{1.1}$$

where $X = (X_1, \ldots, X_n)$ is the input string of length n. The quality of a candidate solution thus improves with the number of 1s in the input string, and the optimum of onemax is in the string of all 1s. To keep things simple, the simulation considers

- a 5-bit onemax,
- a population of size $N = 6$,
- binary tournament selection,

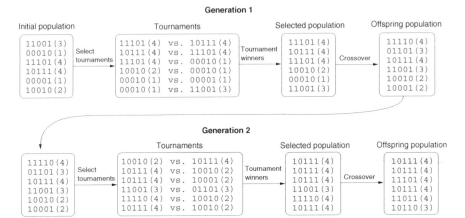

Fig. 1.2. A GA simulation by hand on a 5-bit onemax using a population of size $N = 6$, population-wise uniform crossover, and binary tournament selection. The fitness of each solution is shown in parentheses

- population-wise uniform crossover, and
- no mutation.

Figure 1.2 shows the first two generations of the GA simulation. The initial population of candidate solutions is generated at random. Next, tournament selection is applied to create the set of promising solutions. Six tournaments take place, and the winners of these six tournaments form the population of promising solutions. Population-wise uniform crossover is then applied to the population of selected solutions by shuffling the bits in each position among the selected solutions. The resulting (shuffled) set of solutions then replaces the original population.

In both generations of the simulation, the average fitness of the new population is greater than the average fitness of the population before selection. The fitness increase is good news for us because we want to maximize fitness, but why does this happen? Since for onemax the solutions with more 1s have higher fitness than those with fewer 1s, selection should increase the number of 1s in the population. Crossover neither creates nor destroys any bits in the population. That is why the population after applying selection and crossover should contain more 1s than the original population. Since for onemax fitness increases with the number of 1s, we can expect the overall fitness to increase over time. Ideally, every generation should increase the average fitness of the population unless no improvement is possible.

Nonetheless, the increase of the average fitness tells only half the story. The same increase in the fitness of the population in the first generation would be achieved by using no recombination at all. However, by applying selection without recombination, the best solution in the population would soon overtake the entire population and there would be no chance of finding the real optimum unless this was already in the initial population. Even more importantly, only initial solutions would be considered. Since the initial population in GAs is generated at random, the GA with selection only would perform no better than random search.

Since random search is intractable for problems of moderate or large size, for robust and scalable optimization, we need an additional mechanism besides selection that ensures an effective exploration of new solutions. This can be achieved by combining or mixing superior partial solutions propagated by selection using a recombination operator. From a mixing point of view, population-wise uniform crossover is the most powerful recombination because this recombination operator completely shuffles the bits in each position. Since the proportion of 1s in each position is expected to increase, the probability of finding the optimum by shuffling is expected to increase over time.

We have seen only the first two generations of the simulation. What is the final outcome of the simulation going to look like? In subsequent generations, the selectorecombinative search further increases the quality of candidate solutions. In the simulation, the optimum is found in the third generation. Once the optimum has been found, it quickly (in two more generations) takes over the entire population.

1.4 Population-Wise Uniform Crossover as a Probabilistic Model

Recall that population-wise uniform crossover considers each position independently and creates the new population of candidate solutions by shuffling the bits in each position across the population of promising solutions. Consider using the following *probabilistic* recombination instead:

1. Compute the probability of a 1 in each position in the selected set of promising solutions. The probability of a 1 in position i is denoted by $p(X_i = 1)$ or simply p_i.
2. Generate new candidate solutions by setting the bit in position i of each solution to 1 with probability p_i; otherwise, set the bit to 0. The order in which the bits in each solution are generated does not matter.
3. Repeat step 2 until enough candidate solutions have been generated.

We will refer to the crossover operator described above as *probabilistic uniform crossover*. There is an important difference between population-wise uniform crossover and probabilistic uniform crossover. Population-wise uniform crossover directly manipulates the population of promising solutions to create a new population of candidate solutions. On the other hand, probabilistic uniform crossover first computes single-bit statistics from the selected population, then *discards the population*, and uses only the acquired statistics

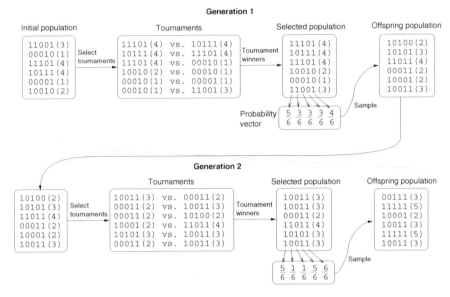

Fig. 1.3. A GA simulation by hand one a 5-bit onemax using a population of size $N = 6$, probabilistic uniform crossover, and binary tournament selection. The fitness of each solution is shown in parentheses

to generate new solutions. Nonetheless, the expected outcome of both operators is the same.

Figure 1.3 shows the first two generations of the simulation using probabilistic uniform crossover instead of population-wise uniform crossover. As discussed above, "recombination" proceeds by computing the probability of 1 in each position of the selected population of promising solutions, and then using the computed probabilities to generate new candidate solutions. Although the proportion of 1s in each position is expected to remain the same after sampling the probability vector, sampling introduces a certain amount of noise; however, for moderate to large populations, the noise becomes negligible. The proportion of 1s continues to increase over time along with the average fitness. With probabilistic uniform crossover, the simulation finds the optimum already in the second generation. After additional two generations, the entire population converges to the optimum.

An example run confirming the growth of the proportion of 1s on a bigger problem is shown in Fig. 1.4, which considers a simulation of probabilistic uniform crossover on a 50-bit onemax. In this figure, the global optimum takes over the entire population after about 20 generations.

It can be shown [74, 118, 175] that the GA with probabilistic uniform crossover converges to the solution of an n-bit onemax with a fixed probability of getting each bit correctly in $O(n)$ fitness evaluations (assuming an adequate population size). If the expected proportion of incorrect bits decreases linearly

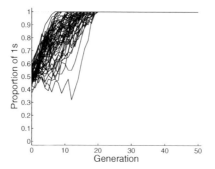

Fig. 1.4. Proportions of 1s on a 50-bit onemax using probabilistic uniform crossover, binary tournament selection, and a population of 100 candidate solutions

with problem size (requiring a constant number of mistaken bits overall), the number of evaluations can be bounded by $O(n \ln n)$.

A near linear growth of the number of function evaluations until reliable convergence to the optimum is good news. Nonetheless, probabilistic uniform crossover ignores the context of each bit (the values of other bits) and that is why its application is limited to problems where the context of each bit does not matter. The following example shows one problem where considering each string position independently is insufficient. Probabilistic uniform crossover is then extended to consider statistics over groups of string positions, ensuring that the global optimum is found scalably.

1.5 Additively Separable Traps and Probabilistic Building-Block Crossover

In an additively separable trap function of order 5 denoted by trap-5, the number of bits in solution strings is assumed to be a multiple of 5. String positions are first partitioned into disjoint partitions of 5 bits each. The partitioning is fixed during the entire optimization run, but it is important that the algorithm is not given information about the string positions in each partition. We denote the positions corresponding to the ith trap partition by indices $b_{i,1}$ to $b_{i,5}$. Trap-5 is then defined as

$$f_{trap,5}(X) = \sum_{i=1}^{\frac{n}{5}} trap_5(X_{b_{i,1}} + X_{b_{i,2}} + \cdots + X_{b_{i,5}}) . \tag{1.2}$$

Each partition $(X_{b_{i,1}}, \ldots, X_{b_{i,5}})$ contributes to the fitness using a trap function of order 5 [1, 38] defined as

$$trap_5(u) = \begin{cases} 5 & \text{if } u = 5 \\ 4 - u & \text{otherwise} \end{cases} , \tag{1.3}$$

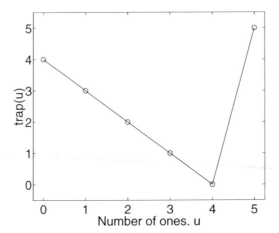

Fig. 1.5. The value of a 5-bit trap function depends on the number of 1s in the input string. The trap has two optima; the global optimum is 11111 and the local optimum is 00000. The average fitness over any subset of bits in the trap leads to the local attractor, unless all the bits are considered together

where u denotes the number of ones in the input block of 5 bits. See Fig. 1.5 to visualize the 5-bit trap. The 5-bit trap function has one global optimum in $u = 5$ and one local optimum in $u = 0$. As a result, an n-bit trap-5 has one global optimum in the string where all bits are equal to 1, and it has $2^{\frac{n}{5}} - 1$ local optima in strings where the bits corresponding to each trap partition are equal, but where the bits in at least one trap partition are equal to 0.

The 5-bit trap is a fully deceptive function [38]; that means that the average fitness of any block of bits of order lower than 5 leads away from the global optimum. For example, the average fitness of a 0 in any position is greater than the average fitness of a 1 in the same position. The same situation can be observed for 00 vs. 11, 000 vs. 111, and 0000 vs. 1111. Since in trap-5, the fitness contribution of each trap is independent of its context, considering blocks of 1, 2, 3, or 4 bits is expected to drive the search away from the global optimum. Figure 1.6a shows the proportions of 1s for different positions using probabilistic uniform crossover on a 50-bit trap-5 consisting of 10 copies of the 5-bit trap. The figure confirms the above argument for single-bit statistics and shows that the proportions of 1s decrease over time, yielding a local optimum in the string of all 0s instead of the global optimum located in the string of all 1s.

Let us assume that the algorithm is capable of discovering where the trap partitions are. It was argued that statistics over strict subsets of positions in each trap partition are misleading. That is why an adequate probabilistic model should consider probabilities of 5-bit partial solutions corresponding to different trap partitions. The probabilistic model should thus estimate probabilities of all instances of 5 bits in each trap partition. New solutions should

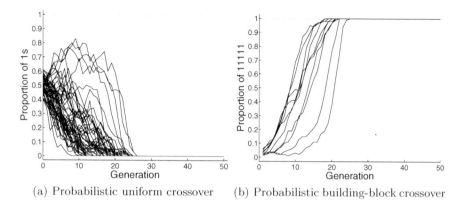

(a) Probabilistic uniform crossover   (b) Probabilistic building-block crossover

Fig. 1.6. Probabilistic uniform crossover and probabilistic building-block crossover on a 50-bit trap-5 consisting of 10 copies of the 5-bit trap using binary tournament selection and a population of 100 candidate solutions

then be generated based on these probabilities, 5 bits at a time. We refer to the probabilistic operator based on statistics over multiple string positions as *probabilistic building-block crossover*.

By considering each trap partition as an intact block, probabilistic building-block crossover is expected to increase the proportion of a 11111 in each trap partition, because the average fitness of a 11111 in any trap partition is expected to be greater than the average fitness of any competitor of this partial solution. Figure 1.6b verifies this argument with an experiment. The figure shows that the optimum takes over the entire population after 23 generations.

Similarly as for onemax, it can be shown that GAs with probabilistic building-block crossover will converge in $O(n)$ evaluations for a fixed probability of getting each trap partition right and $O(n \ln n)$ evaluations for a constant expected number of wrong partitions.

1.6 Building Blocks and Decomposable Problems

Although trap-5 is an artificial example, it suggests that there exist functions for which processing positions independently does not work. For robust and scalable selectorecombinative search, recombination must effectively process those partial solutions that are both (1) contained in the optimum and (2) superior to their competitors over the course of the GA run. We call those superior partial solutions of the optimum *building blocks* [53, 57, 83]. For example, in onemax, a 1 in any position is a building block. On the other hand, in trap-5, a 11111 in any trap partition is a building block.

Since building blocks are superior to their competitors, selection should increase the proportion of the building blocks in the population. If recombination combines promising solutions effectively without disrupting the building blocks, the chances of hitting the optimum are expected to increase over time. Clearly, if building blocks are frequently disrupted, their number can reduce over time thereby reducing the chances of hitting the optimum.

In analyzing scalability of selectorecombinative GAs it is useful to consider classes of problems solvable by considering decomposition into statistics of order bounded by an arbitrary constant k, which also bounds the order of building blocks. We call problems decomposable in this manner *decomposable problems of order k* or *decomposable problems of bounded difficulty* [57].

If recombination combines building blocks contained in solutions effectively (in other words, if recombination considers nonmisleading statistics), the number of evaluations for decomposable problems of bounded difficulty can be bounded by $O(n^2 \ln n)$ [74, 118, 175, 177]. Since many complex real-world problems can be decomposed in some manner, providing low-order polynomial solution for classes of decomposable problems of bounded difficulty seems like a promising approach to solving broad classes of real-world black-box optimization problems.

This reasoning leads to an interesting conclusion: If the order of nonmisleading statistics is bounded by a constant and the probabilistic model considers these statistics, using probabilistic recombination should allow GAs to find the global optimum reliably in only $O(n \ln n)$ to $O(n^2 \ln n)$ evaluations, independently of the order of subproblems. On the other hand, common variation operators – such as one-point crossover, uniform crossover, or bit-flip mutation – yield qualitatively worse performance. More specifically, common crossover operators require exponentially many evaluations to solve arbitrary order-k decomposable problems, whereas mutation requires $O(n^k \log n)$ evaluations. That poses a challenge of how we learn an adequate problem decomposition into nonmisleading statistics. If we can identify the decomposition without a substantial increase in computational cost, probabilistic recombination will provide us with robust and scalable solution for decomposable problems of bounded difficulty.

The following chapter discusses the use of probabilistic recombination in GAs in greater detail and provides a brief overview of algorithms based on this principle.

2

Probabilistic Model-Building Genetic Algorithms

The previous chapter showed that variation operators in genetic and evolutionary algorithms can be replaced by learning a probabilistic model of selected solutions and sampling the model to generate new candidate solutions. Algorithms based on this principle are called probabilistic model-building genetic algorithms (PMBGAs) [133]. This chapter reviews most influential PMBGAs and discusses their strengths and weaknesses. The chapter focuses on PMBGAs working in a discrete domain but other representations are also discussed briefly.

The chapter starts by describing the general PMBGA procedure, which can be seen as a template where one simply fills in methods of choice for learning and sampling probabilistic models. Section 2.2 introduces the family of PMBGAs for optimizing problems in the domain of fixed-length strings over a finite alphabet. Binary strings are used throughout most of the section, but most approaches can be generalized to any finite alphabet in a straightforward manner. Section 2.3 presents some approaches for optimizing problems in other than discrete domains.

2.1 General PMBGA Procedure

The previous chapter presented two probabilistic recombination operators that build and sample a probabilistic model of promising solutions as opposed to using traditional recombination and mutation of GAs. Probabilistic model-building genetic algorithms (PMBGAs) build on this idea and use a probabilistic model of promising solutions to sample new candidate solutions.

The general procedure of PMBGAs is similar to that of GAs (see Sect. 1.2). PMBGAs generate the initial population of candidate solutions at random according to a uniform distribution over the space of all potential solutions. In each iteration, promising solutions are first selected from the current population of candidate solutions. A probabilistic model of the selected solutions is then built and the built probabilistic model is sampled to generate new

Martin Pelikan: *Hierarchical Bayesian Optimization Algorithm*, StudFuzz **170**, 13–30 (2005)
www.springerlink.com

```
Probabilistic model building genetic algorithm (PMBGA)
  t := 0;
  generate initial population P(0);
  while (not done) {
    select population of promising solutions S(t) from P(t);
    build probabilistic model M(t) for S(t);
    sample M(t) to generate new candidate solutions O(t);
    incorporate O(t) into P(t);
    t := t+1;
  };
```

Fig. 2.1. PMBGA pseudocode

candidate solutions. The new solutions are then incorporated into the original population, replacing some of the old ones or all of them. The process is repeated until the termination criteria are met. The basic PMBGA procedure is outlined in Fig. 2.1.

Thus, the difference between PMBGAs and traditional GAs is how these algorithms process populations of promising solutions to generate populations of new candidate solutions. Instead of applying crossover to pairs of selected solutions and then applying mutation to each of the resulting solutions, the following two steps are performed:

1. **Model building.** A probabilistic model of promising solutions is constructed.
2. **Model sampling.** The constructed model is sampled to generate new solutions.

PMBGAs differ in how they cope with the above two steps and in whether they incorporate special selection and replacement mechanisms for processing the populations of solutions. Because building a probabilistic model can be seen as building an estimation of the true probability distribution of promising solutions, PMBGAs are also called *estimation of distribution algorithms* (EDAs) [117], and *iterated density estimation algorithms* (IDEAs) [13].

The search in PMBGAs can also be seen as an iterative procedure for constructing a probabilistic model that generates all global optima. Initially, the probabilistic model encodes a uniform distribution over all potential solutions. Each iteration updates the model to generate better solutions. A successful PMBGA run should end with a probabilistic model that generates all global optima.

The remainder of this chapter provides an overview of most influential PMBGAs. The methods are classified according to the underlying representation of candidate solutions and the complexity of the class of models they employ. The first order of business is to describe PMBGAs that assume that candidate solutions are represented by fixed-length strings over a finite alphabet; we call these algorithms *discrete PMBGAs*. Next, PMBGAs that consider other representations are summarized.

2.2 Discrete Variables

One natural classification of discrete PMBGAs considers the order of interactions that the underlying models employ. PMBGAs have evolved from modeling low-order interactions to modeling interactions of an arbitrary order. This section starts by reviewing PMBGAs that assume that all variables or string positions are independent; these approaches are based on probabilistic uniform crossover discussed earlier. Next, the section discusses PMBGAs capable of covering some pairwise interactions by using probabilistic models in the form of a chain, tree, or forest. The section ends with a brief overview of PMBGAs that can cope with interactions of an arbitrary order. Binary-string representation is assumed, although all discrete PMBGAs can be extended to strings over any finite alphabet in a straightforward manner.

2.2.1 No Interactions

The population-based incremental learning (PBIL) [7] replaces the population of candidate solutions by a *probability vector* $(p_1, p_2, \ldots, p_n)$, where p_i denotes the probability of a 1 in the ith position of solution strings. Each p_i is initially set to 0.5, which corresponds to a uniform distribution over the set of all solutions. In each iteration, PBIL generates s candidate solutions according to the current probability vector where $s \geq 2$ denotes the selection pressure. Each value is generated independently of its context (remaining bits) and thus no interactions are considered (see Fig. 2.2). The best solution from the generated set of s solutions is then used to update the probability-vector entries using

$$p_i = p_i + \lambda(x_i - p_i) \, ,$$

where $\lambda \in (0, 1)$ is the learning rate (say, 0.02), and x_i is the ith bit of the best solution. Using the above update rule, the probability p_i of a 1 in the ith

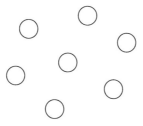

Fig. 2.2. A graphical model with no interactions covered displayed as a Bayesian network. The model with no interactions is equivalent to probabilistic uniform crossover

position increases if the best solution contains 1 in that position and decreases otherwise. In other words, probability-vector entries move *toward* the best solution and, consequently, the probability of generating this solution increases. The process of generating new solutions and updating the probability vector is repeated until some termination criteria are met; for instance, the run can be terminated if all probability-vector entries are close to either 0 or 1.

For example, for a 5-bit onemax, $s = 4$, and $\lambda = 0.02$, the first iteration of PBIL could proceed as follows. Assume that the solutions generated from the initial probability vector $(0.5, 0.5, 0.5, 0.5, 0.5)$ are 10010, 11010, 00101, and 10111. The last solution has the highest fitness and is used to update the probability vector. With $\lambda = 0.02$, the new probability vector is $(0.51, 0.49, 0.51, 0.51, 0.51)$.

PBIL was also referred to as the hill climbing with learning (HCwL) [95] and the incremental univariate marginal distribution algorithm (IUMDA) [113]. Theoretical analyses of PBIL can be found in [63, 81, 95].

The basic difference between PBIL and most GAs (including PMBGAs) is that PBIL does not actually use a population, but it replaces the population by a probability vector. There is a close relation between the learning rate in PBIL and the population size in GAs; decreasing the learning rate λ corresponds to increasing the population size. However, the relation is not fully consistent with the standard GA approach.

The compact genetic algorithm (cGA) [75, 76] eliminates the gap between PBIL and traditional GAs. Like PBIL, cGA replaces the population by a probability vector. All entries in the probability vector are initialized to 0.5. Each iteration updates the probability vector using a variant of binary tournament selection that replaces the worst of the two solutions by the best one using a population of size N. Denoting the bit in the ith position of the best and worst of the two solutions by x_i and y_i, respectively, the probability-vector entries are updated as follows:

$$p_i = \begin{cases} p_i + \frac{1}{N} & \text{if } x_i = 1 \text{ and } y_i = 0 \\ p_i - \frac{1}{N} & \text{if } x_i = 0 \text{ and } y_i = 1 \\ p_i & \text{otherwise} \end{cases} .$$

Although cGA uses a probability vector instead of a population, updates of the probability vector correspond to replacing one candidate solution by another one using a population of size N and shuffling the resulting population using population-wise uniform crossover.

Unlike PBIL and cGA, the univariate marginal distribution algorithm (UMDA) [117] maintains a population of solutions. Each iteration of UMDA starts by selecting a population of promising solutions like traditional GAs. A probability vector is then computed using the selected population of promising solutions and new solutions are generated by sampling the probability vector. The new solutions replace the old ones and the process is repeated until termination criteria are met. UMDA is therefore equivalent to a GA with probabilistic uniform crossover (see Sect. 1.4). Although UMDA uses a probabilistic model as an intermediate step between the original and new populations unlike PBIL and cGA, the performance and dynamics of PBIL, cGA, and UMDA are similar.

PBIL, cGA, and UMDA can solve problems decomposable into subproblems of order one in a linear or quadratic number of fitness evaluations. However, if decomposition into single-bit subproblems misleads the decision making away from the optimum, these algorithms scale up poorly with problem size. For example, PBIL, cGA and UMDA require exponentially many evaluations until reliable convergence for additively separable traps. The next section considers PMBGAs that use models with pairwise dependencies; consequently, these PMBGAs extend the class of problems that can be solved in a scalable manner to problems decomposable into subproblems of order at most two.

2.2.2 Pairwise Interactions

This section focuses on PMBGAs with pairwise probabilistic models that can encode dependencies in the form of a chain, a tree, or a forest (a set of isolated trees). Because these models move beyond the assumption of variable independence, they represent the first step toward competent PMBGAs capable of solving decomposable problems of bounded difficulty in a scalable manner.

The mutual-information-maximizing input clustering (MIMIC) algorithm [33] uses a chain distribution (see Fig. 2.3(a)) specified by (1) an ordering of string positions (variables), (2) a probability of a 1 in the first position of the chain, and (3) conditional probabilities of every other position given the value in the previous position in the chain. A chain probabilistic model encodes the probability distribution where all positions except for the first position of the chain are conditionally dependent on the previous position in the chain.

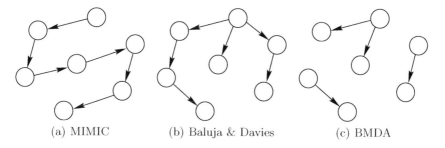

(a) MIMIC   (b) Baluja & Davies   (c) BMDA

Fig. 2.3. Graphical models with pairwise interactions covered displayed as Bayesian networks. The models that can cover some pairwise interactions form the (**a**) chain, (**b**) tree, or (**c**) forest

After selecting promising solutions and computing marginal and conditional probabilities, MIMIC uses a greedy algorithm to maximize mutual information between the adjacent positions in the chain. In this fashion the Kullback-Liebler divergence [94] between the chain and actual distributions is to be minimized. Nonetheless, the greedy algorithm does not guarantee global optimality of the constructed model (with respect to Kullback-Liebler divergence). The greedy algorithm starts in the position with the minimum unconditional entropy. The chain is expanded by adding a new position that minimizes the conditional entropy of the new variable given the last variable in the chain. Once the full chain is constructed for the selected population of promising solutions, new solutions are generated by sampling the distribution encoded by the chain and the probabilities associated with the chain.

There are two important drawbacks of using chain distributions. The first drawback is that chain distributions allow only a very limited representation of dependencies between string positions. Despite that, chain distributions can encode dependencies between pairs of positions that are distant in solution strings; these dependencies are preserved by neither uniform nor one-point crossover. The second drawback is that there is no known algorithm for learning the best chain distribution in polynomial time. Despite the disadvantages of using only chain distributions, the use of pairwise interactions was one of the most important steps in the development of competent PMBGAs capable of solving decomposable problems of bounded difficulty scalably. MIMIC was the first PMBGA to not only learn and use a fixed set of statistics, but it was also capable of identifying the statistics that should be considered to solve the problem efficiently.

Baluja and Davies [8] use dependency trees (see Fig. 2.3(b)) to model promising solutions. Like in PBIL, the population is replaced by a probability vector but in this case the probability vector contains all *pairwise* probabilities. The probabilities are initialized to 0.25. Each iteration adjusts the probability vector according to new promising solutions acquired on the fly. A dependency tree encodes the probability distribution where every variable except for the root is conditioned on the variable's parent in the tree.

A variant of Prim's algorithm for finding the minimum spanning tree [144] can be used to construct an optimal tree distribution. Here the task is to find a tree that maximizes mutual information between parents (nodes with successors) and their children (successors). This can be done by first randomly choosing a variable to form the root of the tree, and "hanging" new variables to the existing tree so that the mutual information between the parent of the new variable and the variable itself is maximized. In this way, the Kullback-Liebler divergence between the tree and actual distributions is minimized as shown by Chow and Liu [24]. Once a full tree is constructed, new solutions are generated according to the distribution encoded by the constructed dependency tree and the conditional probabilities computed from the probability vector.

The bivariate marginal distribution algorithm (BMDA) [139] uses a forest distribution (a set of mutually independent dependency trees, see Fig. 2.3(c)). This class of models is even more general than the class of dependency trees, because any forest that contains two or more disjoint trees cannot be generally represented by a tree. As a measure used to determine whether to connect two variables, BMDA uses a Pearson's chi-square test [104]. This measure is also used to discriminate the remaining dependencies in order to construct the final model. To learn a model, BMDA uses a variant of Prim's algorithm [144].

Pairwise models capture some interactions in a problem with reasonable computational overhead. PMBGAs with pairwise probabilistic models can identify, propagate, and juxtapose partial solutions of order two, and therefore they work well on problems decomposable into subproblems of order at most two [8, 15, 33, 113, 139]. Nonetheless, capturing only some pairwise interactions has still been shown to be insufficient for solving all decomposable problems of bounded difficulty scalably [15, 139]. That is why PMBGA research has pursued more complex models discussed in the next section.

2.2.3 Multivariate Interactions

This section overviews PMBGAs with models that can encode multivariate interactions. Using general multivariate models has brought powerful algorithms capable of solving problems of bounded difficulty quickly, accurately, and reliably. On the other hand, learning distributions with multivariate interactions necessitates more complex model-learning algorithms that require significant computational time and still do not guarantee global optimality of the resulting model. Nonetheless, many difficult problems are intractable using simple models and the use of complex models and algorithms is warranted.

The factorized distribution algorithm (FDA) [116] uses a fixed factorized distribution throughout the whole computation. The model is allowed to contain multivariate marginal and conditional probabilities, but FDA learns only the probabilities, not the structure (dependencies and independencies). To solve a problem using FDA, we must first decompose the problem and then factorize the decomposition. This brings us back to population-wise building-block crossover that is given building-block partitions in advance (see Sect. 1.5). While it is useful to incorporate prior information about the regularities in the search space, the basic idea of black-box optimization is to *learn* the regularities in the search space as opposed to using the regularities specified by the user. In other words, FDA ignores the problem of learning *what statistics* are important to process within the PMBGA framework, and must be given that information in advance.

The extended compact genetic algorithm (ECGA) [70] uses a marginal product model (MPM) that partitions the variables into several partitions, which are processed as independent variables in UMDA (see Fig. 2.4(a)). Each partition is treated as a single variable and different partitions are considered to be mutually independent. The effects of learning and sampling an MPM are

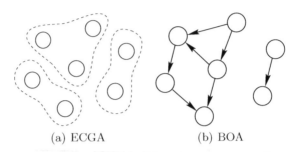

(a) ECGA   (b) BOA

Fig. 2.4. Graphical models with multivariate interactions covered. In ECGA, the variables are partitioned into several groups, each corresponding to one component of the problem decomposition; the partitions are considered independent. In BOA, the variables are related using directed edges, forming a directed acyclic graph called Bayesian network

thus the same as those of probabilistic building-block crossover from Sect. 1.5 with the building-block partitions specified by the MPM.

To decide between alternative MPMs, ECGA uses a variant of the minimum description length (MDL) metric [149, 150, 151], which favors models that allow higher compression of data (in this case, the selected set of promising solutions). More specifically, the Bayesian information criterion (BIC) [164] is used. To find a good model, ECGA uses a greedy algorithm that starts with each variable forming one partition (like in probabilistic uniform crossover). Each iteration of the greedy algorithm merges two partitions that maximize the improvement of the model with respect to BIC. If no more improvement is possible, the current model is used.

ECGA provides robust and scalable solution for problems that can be decomposed into independent subproblems of bounded order (separable problems) [160, 161]. However, many real-world problems contain overlapping dependencies (e.g., two-dimensional spin-glass systems examined in Chap. 7), which cannot be accurately modeled by dividing the variables into disjoint partitions. This can result in poor performance of ECGA on those problems.

The Bayesian optimization algorithm (BOA) [129] builds a Bayesian network for the population of promising solutions (see Fig. 2.4(b)) and samples the built network to generate new candidate solutions. Initially, BOA used the Bayesian-Dirichlet metric subject to a maximum model-complexity constraint [27, 79] to discriminate competing models, but other metrics have been analyzed in later work. In all variants of BOA, the model is constructed by a greedy algorithm that iteratively adds a new dependency in the model that maximizes the model quality. Other elementary graph operators – such as edge removals and reversals – can be incorporated, but edge additions are most important. The construction is terminated when no more improvement is possible. The greedy algorithm used to learn a model in BOA is similar to the one used in ECGA. However, Bayesian networks can encode more complex dependencies and independencies than models used in ECGA. Therefore, BOA is also applicable to problems with overlapping dependencies. Several such problems will be discussed in Chap. 7.

BOA uses an equivalent class of models as FDA; however, BOA learns both the structure and the probabilities of the model. Although BOA does not require problem-specific knowledge in advance, prior information about the problem can be incorporated using Bayesian statistics, and the relative influence of prior information and the population of promising solutions can be tuned by the user. An interesting study on incorporating prior problem-specific information to improve the performance of BOA in graph partitioning can be found in Schwarz and Ocenasek [166].

A discussion of the use of Bayesian networks as an extension to tree models can also be found in Baluja and Davies [9]. An algorithm that uses Bayesian networks to model promising solutions was independently developed by Etxeberria and Larrañaga [41], who called it the estimation of Bayesian network algorithm (EBNA). Mühlenbein and Mahnig [115] later improved the original FDA by using Bayesian networks together with the greedy algorithm for learning the networks described above. The modification of FDA was named the learning factorized distribution algorithm (LFDA).

Helmholtz machines used in the Bayesian evolutionary algorithm proposed by Zhang and Shin [183] can also encode multivariate interactions. Helmholtz machines encode interactions by introducing new, hidden variables, which are connected to every variable.

PMBGAs that use models capable of covering multivariate interactions can solve a wide range of problems in a scalable manner; promising results were reported on two-dimensional Ising spin-glass systems [114, 129], graph partitioning [165, 166], telecommunication network optimization [154], silicon cluster optimization [160], and scheduling [100]. Later (see Chaps. 5 and 6) we will see how hierarchical decomposition can be used to further extend the applicability of BOA beyond problems decomposable into tractable subproblems on a single level.

2.3 Other Representations

There are two basic approaches to extending PMBGAs for discrete fixed-length strings to other domains such as variable-length strings, vectors of real numbers, symbolic expressions, and program codes:

1. Map the problem to the domain of fixed-length discrete strings, solve the discrete problem, and map the solution back to the problem's original domain.
2. Extend or modify the class of probabilistic models to other domains.

The first approach has been studied in the context of genetic and evolutionary algorithms for several decades. This section reviews PMBGAs based on the second approach. The section starts with an overview of PMBGAs for optimizing problems over the fixed-length vectors of real-valued variables. Next, the section discusses PMBGAs for optimizing computer programs and symbolic expressions.

2.3.1 Real-Valued Variables

There are many approaches to modeling and sampling real-valued distributions. One way of classifying these approaches considers the number of variables that are treated together; some approaches build a separate model for each variable (like PBIL), whereas other approaches model groups of variables or all variables together (like ECGA or BOA). Another way of classifying real-valued PMBGAs considers the type of distributions that are used to model each variable or each group of variables; real-valued models include normal distributions, joint normal distributions, histogram distributions, uniform distributions over intervals, mixture distributions, and others. This section starts by presenting PMBGAs based on normal distributions. Next, the section discusses real-valued PMBGAs based on histogram and interval distributions.

Single-Peak Normal Distributions

The stochastic hill climbing with learning by vectors of normal distributions (SHCLVND) [155] is a straightforward extension of PBIL to vectors of real-valued variables using a normal distribution to model each variable. SHCLVND replaces the population of real-valued solutions by a vector of means $\mu = (\mu_1, \ldots, \mu_n)$, where μ_i denotes a mean of the distribution for the ith variable. The same standard deviation σ is used for all variables. See Fig. 2.5 for an example model. At each generation, a random set of solutions is first generated according to μ and σ. The best solution out of this subset is then used to update the entries in μ by shifting each μ_i toward the value of ith variable in the best solution using an update rule similar to the one used in PBIL. Additionally, each generation reduces the standard deviation to make the future exploration of the search space narrower. A similar algorithm was independently developed by Sebag and Ducoulombier [168], who also discussed several approaches to evolving a standard deviation for each variable.

Mixtures of Normal Distributions

The probability density function of a normal distribution is centered around its mean and decreases exponentially with the distance from the mean. If there are multiple "clouds" of values, a normal distribution must either focus on only one of these clouds, or it can embrace multiple clouds at the expense of including the area between these clouds. In both cases, the resulting distribution cannot model the data accurately. One way of extending standard

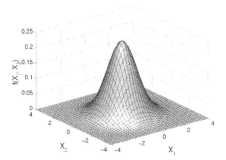

Fig. 2.5. The product of one-dimensional Gaussian distributions as used in the stochastic hill-climbing with learning by vectors of normal distributions (SHCLVND) of Rudlof et al. (1996). Each variable is modeled by a normal distribution, and the overall distribution is given by the product of the individual distributions for each variable

single-peak normal-distribution models to enable coverage of multiple groups of similar points is to use a mixture of normal distributions. Each component of the mixture of normal distributions is a normal distribution by itself. A coefficient is specified for each component of the mixture to denote the probability that a random point belongs to this component. The probability density function of a mixture is thus computed by multiplying the density function of each mixture component by the probability that a random point belongs to the component, and adding these weighted densities together.

Gallagher et al. [46] extended PMBGAs based on single-peak normal distributions by using an adaptive mixture of normal distributions to model each variable. The parameters of the mixture (including the number of components) evolve based on the discovered promising solutions. Using mixture distributions is a significant improvement compared to single-peak normal distributions, because mixtures allow simultaneous exploration of multiple basins of attraction for each variable.

Within the IDEA framework, Bosman and Thierens [13] proposed IDEAs using the joint normal *kernels* distribution, where a single normal distribution is placed around each selected solution (see Fig. 2.7). A joint normal kernels distribution can be therefore seen as an extreme use of mixture distributions with one mixture component per point in the training sample. The variance of each normal distribution can be either fixed to a relatively small value, but it should be preferable to adapt variances to the current state of search. Using kernel distributions corresponds to using a fixed zero-mean normally distributed mutation for each promising solution as is often done in evolution strategies [146]. That is why it is possible to directly take up strategies for adapting the variance of each kernel from evolution strategies [69, 146, 147, 167].

Joint Normal Distributions and Their Mixtures

What changes when instead of fitting each variable with a separate normal distribution or a mixture of normal distributions, groups of variables are considered together? Let us first consider using a single-peak normal distribution. In multivariate domains, a joint normal distribution can be defined by a vector of n means (one mean per variable) and a *covariance matrix* of size $n \times n$. Diagonal elements of the covariance matrix specify the variances for all variables, whereas nondiagonal elements specify linear dependencies between pairs of variables. Considering each variable separately corresponds to setting all nondiagonal elements in a covariance matrix to 0. Using different deviations for different variables allows for "squeezing" or "stretching" the distribution along the axes. On the other hand, using nondiagonal entries in the covariance matrix allows rotating the distribution around its mean. Figure 2.6 illustrates the difference between a joint normal distribution using only diagonal elements of the covariance matrix and a distribution using the full covariance matrix. Therefore, using a covariance matrix introduces another degree of freedom and improves the expressiveness of a distribution. Again, one can use a number of joint normal distributions in a mixture, where each component consists of its mean, covariance matrix, and weight.

A joint normal distribution including a full or partial covariance matrix was used within the IDEA framework [13] and in the estimation of Gaussian networks algorithm (EGNA) [97]. Both these algorithms can be seen as extensions of PMBGAs that model each variable by a single normal distribution to use also nondiagonal elements of the covariance matrix.

Bosman and Thierens [14] proposed *mixed IDEAs* as an extension of PMB-GAs that use a mixture of normal distributions to model each variable. Mixed

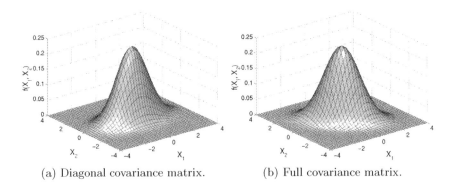

(a) Diagonal covariance matrix.   (b) Full covariance matrix.

Fig. 2.6. A joint normal distribution. The *left-hand* side shows a joint normal distribution with only diagonal entries in the covariance matrix; all remaining entries in the covariance matrix are set to 0. The *right-hand* side shows a joint normal distribution that includes also the covariance between pairs of variables, which allows rotating the distribution around its mean

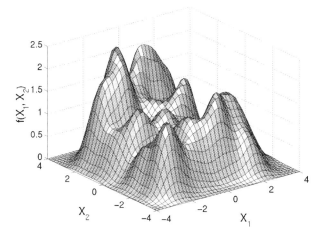

Fig. 2.7. A joint normal kernels distribution is a mixture of joint normal distributions, where a special mixture component is reserved for each point in the modeled data

IDEAs allow multiple variables to be modeled by a separate mixture of joint normal distributions. At one extreme, each variable can have a separate mixture; at another extreme, one mixture of joint distributions covering all the variables is used. Despite that learning such a general class of distributions is quite difficult and a large number of samples is necessary for reasonable accuracy, good results were reported on single-objective [14] as well as multiobjective problems [90, 98, 174]. Using mixture models for all variables was also proposed as a technique for reducing model complexity in discrete PMBGAs [125].

Real-valued PMBGAs presented so far are applicable to real-valued optimization problems without requiring differentiability or continuity of the underlying problem. However, if it is possible to at least partially differentiate the problem, gradient information can be used to incorporate some form of gradient-based local search and the performance of real-valued PMBGAs can be significantly improved. A study on combining real-valued PMBGAs within the IDEA framework with gradient-based local search can be found in Bosman and Thierens [12].

To summarize, the following questions are important in the design of real-valued PMBGAs based on normal distributions or mixtures of normal distributions:

- **Decomposition.** Use a separate distribution estimate for each variable, or consider groups of variables together?
- **Covariances.** If considering multiple variables together, use a diagonal, partial, or full covariance matrix?

- **Mixtures.** Use one-peak normal distribution, or a mixture of normal distributions?
- **Kernels.** Use a mixture of normal distributions with one component per solution (joint normal kernels distribution), or attempt to create a global model that generalizes the data globally?

Different models might be advantageous for different types of problems. For successful application of real-valued PMBGAs based on normal distributions it is important to consider the above questions and make decisions based on the properties of the problem. For example, if the problem is highly multimodal, using a single-peak distribution will most likely not work; however, if there are extremely many local optima, using single-peak distributions could generalize the problem landscape and avoid getting stuck in a local optimum. Furthermore, it is important to consider the choice of a distribution to determine an adequate population size and other parameters. For example, while for joint normal kernels distributions populations of only few points might suffice, for adapting complex multivariate mixture distributions a rather large sample is required.

Other Real-Valued PMBGAs

Of course, using normal distributions is not the only approach to modeling real-valued distributions. Other density functions are frequently used to model real-valued probability distributions, including histogram distributions, interval distributions, and others. A brief review of real-valued PMBGAs that use other than normal distributions follows.

In the algorithm proposed by Servet et al. [170], an interval (a_i, b_i) and a number $z_i \in (0, 1)$ are stored for each variable (see Fig. 2.8). By z_i, the probability that the ith variable is in the lower half of (a_i, b_i) is denoted. Each z_i is initialized to 0.5. To generate a new candidate solution, the value of each variable is selected randomly from the corresponding interval. The

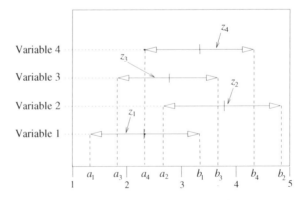

Fig. 2.8. A model based on adaptive intervals (Servet et al., 1998)

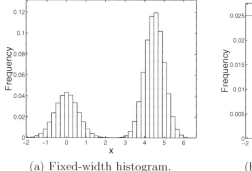

(a) Fixed-width histogram.

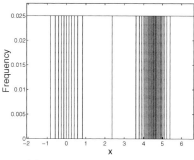

(b) Fixed-height histogram.

Fig. 2.9. Fixed-width and fixed-height histograms. In a fixed-width histogram, each bin has equal width, and the height of each bin determines the number of points in that bin. In a fixed-height histogram, the width of each bin is set so that each bin contains the same number of points

best solution is then used to update the value of each z_i. If the value of the ith variable of the best solution is in a lower half of (a_i, b_i), z_i is shifted toward 0; otherwise, z_i is shifted toward 1. When z_i gets close to 0, interval (a_i, b_i) is reduced to its lower half; if z_i gets close to 1, interval (a_i, b_i) is reduced to its upper half. Figure 2.8 shows an example probabilistic model based on adaptive intervals; in the figure, each z_i is mapped to the corresponding interval (a_i, b_i).

Bosman and Thierens [13], Tsutsui et al. [178] and Cantú-Paz [17] use empirical histograms to model each variable as opposed to using a single normal distribution or a mixture of normal distributions. In these approaches, a histogram for each single variable is constructed. New points are then generated according to the distribution encoded by the histograms for all variables. Figure 2.9 shows examples of fixed-height and fixed-width histograms. The sampling of a histogram proceeds by first selecting a particular bin based on its relative frequency, and then generating a random point from the interval corresponding to the bin. It is straightforward to replace the histograms in the above methods by various classification and discretization methods of statistics and machine learning (such as k-means clustering).

Pelikan et al. [135, 136] use an adaptive mapping from the continuous domain to the discrete one in combination with discrete PMBGAs. The population of promising solutions is first discretized using equal-width histograms, equal-height histograms, k-means clustering, or other classification techniques. A population of promising discrete solutions is then selected. New points are created by applying a discrete recombination operator to the selected population of promising discrete solutions. For example, new solutions can be generated by building and sampling a Bayesian network like in BOA. The resulting discrete solutions are then mapped back into the continuous domain by sampling each class (a bin or a cluster) using the original values of the variables in the selected population of continuous solutions (before discretization). The resulting solutions are perturbed using one of the adaptive mutation operators of evolution strategies [69, 146, 147, 167]. In this way, competent discrete PMBGAs can be combined with advanced methods based on adaptive local search in the continuous domain.

The mixed Bayesian optimization algorithm (mBOA) developed by Ocenasek and Schwarz [121] models vectors of continuous variables with an extension of Bayesian networks with local structures. A model used in mBOA consists of a decision tree for each variable. Each internal node in the decision tree for a variable is a test on the value of another variable. Each test on a variable is specified by a particular value, which is also included in the node. The test considers two cases: the value of the variable is greater or equal than the value in the node or it is smaller. Each internal node has two children, each child corresponding to one of the two results of the test specified in this node. Leaves in a decision tree thus correspond to rectangular regions in the search space. For each leaf, the decision tree for the variable specifies a single-variable mixture of normal distributions centered around the values of this variable in the solutions consistent with the path to the leaf. Thus, for each variable, the model in mBOA divides the space reduced to other variables into rectangular regions, and it uses a single-variable normal kernels distribution to model the variable in each region.

2.3.2 Computer Programs

Genetic programming [92, 93] evolves a population of computer programs using variants of crossover and mutation appropriate to a program code. Programs are represented by labeled trees. Internal nodes of a program tree correspond to functions with at least one argument, the successors (children) of an internal node represent arguments of the function encoded in the node, and leaves of the program tree corresponds to terminal symbols (functions with no arguments, variables, or constants).

Since the context of different parts of a program usually matters a great deal, it should be advantageous to use competent PMBGAs capable of identifying an adequate problem decomposition in genetic programming. Using PMBGAs in genetic programming poses two challenges: (1) computer programs are not linear structures, and (2) the length of computer programs can vary. The remainder of this section presents several approaches to applying PMBGAs to genetic programming.

The probabilistic incremental program evolution (PIPE) algorithm [158, 159] uses a probabilistic model in the form of a tree of a specified maximum allowable size. Nodes in the model specify the probabilities of functions and terminals. PIPE does not employ any interactions among the nodes in the model. To visualize the model used in PIPE, see Fig. 2.10. The model is updated by adapting the probabilities based on the population of selected solutions. New program trees are generated in a top-down fashion starting in

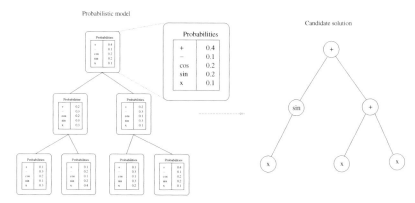

Fig. 2.10. This figure shows an example probabilistic model of a program with no interactions covered used in PIPE and a candidate solution (mathematic expression) generated by the model. Each node in the model stores the probabilities of functions and terminal symbols. The sampling proceeds recursively in a top-down fashion. If a terminal is generated in a node, the generation along this branch is terminated. If a function with one or more arguments is generated, the generation continues by generating the subtrees corresponding to the arguments of the function

the root and continuing to lower levels of the tree if necessary. More specifically, if the model generates a function in a node and that function requires additional arguments, the successors (children) of the node are generated to form the arguments of the function. If a terminal is generated, the generation along the considered branch terminates.

A hierarchical extension of PIPE – the so-called H-PIPE – has been later proposed by Salustowicz and Schmidhuber [157]. In H-PIPE, nodes of a model are allowed to contain subroutines. Both the subroutines and the overall program are evolved.

Handley [68] uses tree probabilistic models to represent the population of programs (trees) in genetic programming. Although the goal of this work was to compress the population of computer programs in genetic programming, Handley's approach can be used within the PMBGA framework to model and sample candidate solutions represented by computer programs or symbolic expressions.

The extended compact genetic programming (ECGP) [162] considers a maximum tree of maximum branching like PIPE. Nonetheless, ECGP uses a marginal product model similar to that used in ECGA, where the nodes in the model are partitioned into clusters of variables that strongly depend on each other.

While ECGP extends ECGA to the domain of computer programs, Looks et al. [102] extend BOA to the same domain. Combinatory logic is used to represent program trees in a unified manner. Program trees translated with combinatory logic are then modeled with Bayesian networks like in BOA.

Contrary to other approaches to GP presented in this section, here complexity of computer programs is not limited, but solutions are allowed to become more complex if necessary.

The following chapter returns to the domain of fixed-length strings over a finite alphabet, which is the main focus of this book. The Bayesian optimization algorithm (BOA) is described, which uses Bayesian networks to model promising solutions found so far and to sample new candidate solutions.

3

Bayesian Optimization Algorithm

The previous chapter argued that using probabilistic models with multivariate interactions is a powerful approach to solving problems of bounded difficulty. The Bayesian optimization algorithm (BOA) combines the idea of using probabilistic models to guide optimization and the methods for learning and sampling Bayesian networks. To learn an adequate decomposition of the problem, BOA builds a Bayesian network for the set of promising solutions. New candidate solutions are generated by sampling the built network.

The purpose of this chapter is threefold. First, the chapter describes the Bayesian optimization algorithm (BOA), which uses Bayesian networks to model promising solutions and bias the sampling of new candidate solutions. Second, the chapter describes how to learn and sample Bayesian networks. Third, the chapter tests BOA on a number of challenging decomposable problems.

The chapter starts by presenting the basic procedure of BOA. Section 3.2 describes Bayesian networks. Section 3.3 discusses how to learn the structure and parameters of Bayesian networks given a sample from an unknown target distribution. The section provides two approaches to measuring the quality of each candidate model: (1) Bayesian metrics and (2) minimum description length metrics. Additionally, the section describes a greedy algorithm for learning the structure of Bayesian networks. Section 3.4 describes how to sample a Bayesian network to generate new candidate solutions. Section 3.5 closes the chapter by presenting initial experiments indicating good scalability of BOA on problems of bounded difficulty.

3.1 Description of BOA

The Bayesian optimization algorithm (BOA) [129, 130, 132] evolves a population of candidate solutions by building and sampling Bayesian networks. BOA can be applied to black-box optimization problems where candidate solutions

Martin Pelikan: *Hierarchical Bayesian Optimization Algorithm*, StudFuzz **170**, 31–48 (2005)
www.springerlink.com

```
Bayesian optimization algorithm (BOA)
  t := 0;
  generate initial population P(0);
  while (not done) {
    select population of promising solutions S(t);
    build Bayesian network B(t) for S(t);
    sample B(t) to generate O(t);
    incorporate O(t) into P(t);
    t := t+1;
  };
```

Fig. 3.1. The pseudocode of the Bayesian optimization algorithm (BOA)

are represented by fixed-length strings over a finite alphabet, but for the sake of simplicity, only binary strings are considered in most of this chapter.

BOA generates the initial population of strings at random with a uniform distribution over all possible strings. The population is updated for a number of iterations (generations), each consisting of four steps. First, promising solutions are selected from the current population using a GA selection method, such as tournament or truncation selection. Second, a Bayesian network that fits the population of promising solutions is constructed. Third, new candidate solutions are generated by sampling the built Bayesian network. Fourth, the new candidate solutions are incorporated into the original population, replacing some of the old ones or all of them.

The above four steps are repeated until some termination criteria are met. For instance, the run can be terminated when the population converges to a singleton, the population contains a good enough solution, or a bound on the number of iterations has been reached. The basic BOA procedure is outlined in Fig. 3.1.

There are a number of alternative ways to perform each step. The initial population can be biased according to prior problem-specific knowledge [160, 166]. Selection can be performed using any popular selection method. Various algorithms can be used to construct the model and there are several approaches to evaluating quality of candidate models. Additionally, the measure of model quality can incorporate prior information about the problem to enhance the estimation and, as a consequence, to improve the efficiency.

The next section describes Bayesian networks and discusses techniques for learning and sampling Bayesian networks.

3.2 Bayesian Networks

A Bayesian network [88, 123] is defined by two components:

(1) **Structure.** The structure is encoded by a directed acyclic graph with the nodes corresponding to the variables in the modeled data set (in this case, to the positions in solution strings) and the edges corresponding to conditional dependencies.

(2) **Parameters.** The parameters are represented by a set of conditional probability tables specifying a conditional probability for each variable given any instance of the variables that the variable depends on.

Mathematically, a Bayesian network encodes a joint probability distribution given by

$$p(X) = \prod_{i=1}^{n} p(X_i | \Pi_i) , \tag{3.1}$$

where $X = (X_1, \ldots, X_n)$ is a vector of all the variables in the problem; Π_i is the set of parents of X_i in the network (the set of nodes from which there exists an edge to X_i); and $p(X_i | \Pi_i)$ is the conditional probability of X_i given its parents Π_i.

A directed edge relates the variables so that in the encoded distribution, the variable corresponding to the terminal node is conditioned on the variable corresponding to the initial node. More incoming edges into a node result in a conditional probability of the variable with the condition consisting of all parents of this variable. In addition to encoding dependencies, each Bayesian network encodes a set of independence assumptions. Independence assumptions state that each variable is independent of any of its antecedents in the ancestral ordering, given the values of the variable's parents.

A simple example Bayesian network structure is shown in Fig. 3.2. The example network encodes a number of conditional dependencies. For instance, the speed of the car depends on whether it is raining and/or radar is enforced. The road is most likely wet if it is raining. Additionally, the network encodes a number of simple and conditional independence assumptions. For instance, the radar enforcement is independent of whether it is raining or not. A more complex conditional independence assumption is that the probability of an accident is independent of whether the radar is enforced, given a particular speed and whether the road is wet. To fully specify the Bayesian network

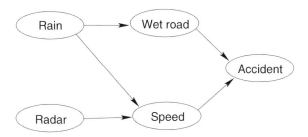

Fig. 3.2. An example Bayesian network structure

Table 3.1. If we encode the speed of a car using two values (high and low), the conditional probability table for the probability of an accident could be defined by this table. Note that the last four entries in the table are unnecessary, because they can be computed using the remaining ones (the sum of all conditional probabilities with a fixed condition is 1)

Accident	Wet Road	Speed	$p(\text{Accident}|\text{Wet Road}, \text{Speed})$
Yes	Yes	High	0.18
Yes	Yes	Low	0.04
Yes	No	High	0.06
Yes	No	Low	0.01
No	Yes	High	0.82
No	Yes	Low	0.96
No	No	High	0.94
No	No	Low	0.99

with the structure shown in the figure, it would be necessary to add a table of conditional probabilities for each variable. An example conditional probability table is shown in Table 3.1.

It is important to understand the semantics of Bayesian networks in the framework of PMBGAs. Conditional *dependencies* will cause the involved variables to remain in the configurations seen in the selected population of promising solutions. On the other hand, conditional *independencies* lead to the mixing of bits and pieces of promising solutions in some contexts (the contexts are determined by the variables in the condition of the independency). The complexity of a proper model is directly related to a proper problem decomposition discussed in Chap. 1. If the problem was linear, a good network would be the one with no edges (see Fig. 3.3(a)); the effects of using an empty network are the same as those of using population-wise uniform crossover. On the other hand, if the problem consisted of traps of order k, the network should be composed of fully connected sets of k nodes, each corresponding to one trap, with no edges between the different groups (see Fig. 3.3(b)); the

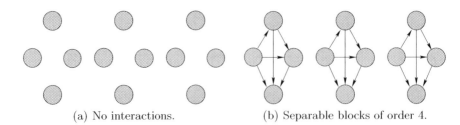

(a) No interactions.   (b) Separable blocks of order 4.

Fig. 3.3. A good model for a problem with no interactions and a problem consisting of separable subproblems of order 4

effects of using such a model would be the same as those of population-wise building-block crossover. More complex problems lead to more complex models, although many problems can be solved with quite simple networks despite the presence of nonlinear interactions of high order.

The following section discusses how to learn a Bayesian network given a data set. Subsequently, we describe the probabilistic logic sampling, which can be used to sample the distribution encoded by a Bayesian network and its parameters.

3.3 Learning Bayesian Networks

For a successful application of BOA, it is necessary that BOA is capable of *learning* a network that reflects the dependencies and independencies that decompose the problem properly. There are two subtasks of learning a Bayesian network:

(1) **Learning the structure.** First, the structure of a network must be determined. The structure defines conditional dependencies and independencies encoded by the network.

(2) **Learning the conditional probabilities.** The structure also identifies conditional probabilities that must be specified for a complete model. After learning the structure, the values of the conditional probabilities with respect to the final structure must be learned.

In BOA, learning the parameters for a given structure is simple, because the value of each variable in the population of promising solutions is specified; in other words, we can assume complete data. To maximize the likelihood of the model with a fixed structure and complete data, the probabilities should be set according to the relative frequencies observed in the modeled data (in our case, the selected set of promising solutions) [79]. Thus, the parameters can be learned by iterating through all selected solutions and computing relative frequencies of different partial solutions.

As an example of learning the parameters of a Bayesian network, recall probabilistic uniform crossover, which can be represented by a Bayesian network with no edges. The parameters of an empty network consist of the probabilities of different values for each variable. Therefore, to determine the parameters of an empty network, it is sufficient to parse the population of promising solutions and compute the observed probabilities.

Learning the structure is a much more difficult problem. It is common to split the problem of learning the structure of Bayesian networks into two components:

1. **A scoring metric.** The scoring metric measures quality of Bayesian network structures. In GA terminology, the scoring metric specifies the fitness of a structure. Usually, the scoring metric is proportional to the likelihood

of the structure or it is equal to a combination of the likelihood and a penalty that grows with model complexity. However, other measures can be used, such as statistical tests on independence.

2. **A search procedure.** The search procedure searches the space of all possible network structures to find the best network with respect to a given scoring metric. The space of network structures can be restricted according to a bound on the complexity of networks or some other prior problem-specific knowledge.

Next, the section discusses several approaches to evaluating competing network structures. Subsequently, the section describes a greedy algorithm, which can be used to learn the structure of a Bayesian network given a scoring metric.

3.3.1 Scoring Metric

There are two approaches to measuring quality of competing network structures: (1) Bayesian metrics and (2) minimum description length metrics. Bayesian metrics [27, 79] measure the quality of each structure by computing the marginal likelihood of the structure with respect to the given data and inherent uncertainties. Minimum description length metrics [149, 150, 151] are based on the assumption that model quality is proportional to the amount of compression of the data allowed by the model.

Bayesian Metrics

Measuring the quality of a network structure is difficult because there is no information about the target network structure or its parameters. Bayesian metrics [27, 79] account for these sources of uncertainty by using Bayes rule and assigning prior distributions to both network structures as well as the parameters of each structure. The quality of a structure is measured by the marginal likelihood of the structure with respect to the given data. The marginal likelihood is computed by averaging the likelihood of the models conditioned on the observed data according to a prior distribution over all possible conditional probabilities in the model:

$$p(B|D) = \frac{p(B)}{p(D)} \int_\theta p(\theta|B)p(D|B,\theta) \, d\theta \ , \tag{3.2}$$

where B is the evaluated Bayesian network structure (without parameters); D is the data set; and each value of θ represents one possible way of assigning conditional probabilities in the network B. Furthermore, $p(B)$ is the prior probability of the network structure B, $p(\theta|B)$ is the prior probability of parameters θ (conditional probabilities) given B, and $p(D|B,\theta)$ denotes the probability of D given the network structure B and its parameters θ. Since the probability of data denoted in the last equation by $p(D)$ is the same for all network structures, $p(D)$ is usually omitted when evaluating the structures.

To compute the marginal likelihood, a prior probability distribution over the parameters of each structure must be given. The Bayesian-Dirichlet metric (BD) [27, 79] assumes that the conditional probabilities follow Dirichlet distribution and makes a number of additional assumptions, yielding the following score:

$$BD(B) = p(B) \prod_{i=1}^{n} \prod_{\pi_i} \frac{\Gamma(m'(\pi_i))}{\Gamma(m'(\pi_i) + m(\pi_i))} \prod_{x_i} \frac{\Gamma(m'(x_i, \pi_i) + m(x_i, \pi_i))}{\Gamma(m'(x_i, \pi_i))},$$
(3.3)

where $p(B)$ is the prior probability of the network structure B; the product over x_i runs over all instances of x_i (in the binary case these are 0 and 1); the product over π_i runs over all instances of the parents Π_i of X_i (all possible combinations of values of Π_i); $m(\pi_i)$ is the number of instances with the parents Π_i set to the particular values given by π_i; and $m(x_i, \pi_i)$ is the number of instances with $X_i = x_i$ and $\Pi_i = \pi_i$. Terms $m'(\pi_i)$ and $m'(x_i, \pi_i)$ denote prior information about the statistics $m(\pi_i)$ and $m(x_i, \pi_i)$, respectively. Here we consider K2 metric [27], which uses an uninformative prior that assigns $m'(x_i, \pi_i) = 1$ and $m'(\pi_i) = \sum_{x_i} m'(x_i, \pi_i)$.

A prior distribution over network structures specified by term $p(B)$ can bias the construction toward particular structures by assigning higher prior probabilities to those preferred structures. Prior knowledge about the structure permits the assignment of higher prior probabilities to those networks similar to the structure believed to be close to the correct one [79]. The search can also be biased toward simpler models by assigning higher prior probabilities to models with fewer edges or parameters [23, 44, 134]. If there is no prior information about the network structure, the probabilities $p(B)$ are set to a constant and omitted in the construction (uniform prior).

For further details on Bayesian metrics, please refer to Cooper and Herskovits [27] and Heckerman et al. [79].

Minimum Description Length Metrics

Minimum description length metrics [149, 150, 151] are based on the assumption that model quality is somehow proportional to the amount of compression of the data allowed by the model. A model that results in the highest compression should capture most inherent regularities. There are two major approaches to the design of MDL metrics. The first is based on a two-part coding where the score is negatively proportional to the sum of the number of bits required to store (1) the model and (2) the data compressed according to the model. The second approach is based on universal code, which normalizes the conditional probability of data given a model by the sum of the probabilities of all data sequences given the model. The normalized probability of the data is used as the basis for computing the number of bits required to compress the data.

In this work we consider one two-part MDL metric called the Bayesian information criterion (BIC) [164] used previously in the extended compact genetic algorithm (ECGA) [70] and the estimation of Bayesian networks algorithm (EBNA) [96]. In the binary case, BIC assigns the network structure a score

$$BIC(B) = \sum_{i=1}^{n} \left(-H(X_i|\Pi_i)N - 2^{|\Pi_i|}\frac{\log_2(N)}{2} \right), \qquad (3.4)$$

where $H(X_i|\Pi_i)$ is the conditional entropy of X_i given its parents Π_i; n is the number of variables; and N is the population size (the size of the training data set). The conditional entropy $H(X_i|\Pi_i)$ is given by

$$H(X_i|\Pi_i) = -\sum_{x_i,\pi_i} p(x_i,\pi_i)\log_2 p(x_i|\pi_i), \qquad (3.5)$$

where $p(x_i, \pi_i)$ is the probability of instances with $X_i = x_i$ and $\Pi_i = \pi_i$; and $p(x_i|\pi_i)$ is the conditional probability of instances with $X_i = x_i$ given that $\Pi_i = \pi_i$.

$H(X_i|\Pi_i)$ denotes the average number of bits required to store a value of X_i given a value of Π_i. BIC multiplies the entropy $H(X_i|\Pi_i)$ by the population size N to reflect the number of bits required to store the entire population. The term $\log_2(N)$ denotes the number of bits required to store one parameter of the model (one probability or frequency). The number of bits required to store each parameter is divided by two because only half of the bits really matter in practice [45]. The term with the conditional entropy ensures that the more the information about the parents of a variable enables to compress the values of the variable, the higher the value of the BIC metric. The term with the $\log_2(N)$ introduces the pressure toward simpler models by decreasing the metric in proportion to the number of parameters required to fully specify the network.

For further details on the minimum description length metrics, please see Rissanen [151] and Grünwald [67].

According to our experience, Bayesian metrics tend to be too sensitive to noise in the data and often capture unnecessary dependencies. To avoid overly complex models, the space of network structures must usually be restricted by specifying a maximum order of interactions [79, 132]. On the other hand, MDL metrics favor simple models so that no unnecessary dependencies need to be considered. In fact, MDL metrics often result in overly simple models and require large populations to learn a model that captures all necessary dependencies.

3.3.2 Search Procedure

Learning the structure of a network given a scoring metric is a difficult combinatorial problem. It has been shown that finding the best network is

NP-complete for most Bayesian and non-Bayesian metrics [22]; therefore, there is no known polynomial-time algorithm for finding the best network structure with respect to most scoring metrics. However, a simple greedy algorithm [79] often performs well and has been successfully used in a number of difficult machine learning tasks.

The greedy algorithm performs an elementary graph operation that improves the quality of the current network the most until no more improvement is possible. The network structure can be initialized to the graph with no edges or the best tree graph computed using the polynomial-time maximum branching algorithm [39]. In BOA, the initial structure can be also set to the structure learned in the previous generation. In all experiments presented in this book, the network is constructed from an empty network in every generation. The following three elementary operations are commonly used:

1. **Edge addition.** An edge is added into the network to add a new dependency.
2. **Edge removal.** An existing edge is removed from the current network to remove an existing dependency, and introduce a new independence assumption or make an existing independence assumption stronger.
3. **Edge reversal.** An existing edge is reversed to change the character of the corresponding dependency. Each edge reversal can be replaced by first removing the edge and then adding the reversed edge in its place.

The search is terminated when there is no operation that can improve the score of the current metric. It is necessary to ensure that after each performed operation on the network structure, the resulting graph represents a valid Bayesian network structure. Consequently, the operations that introduce cycles in the structure must be eliminated. Additionally, it is useful to limit the number of edges that end in any node to upper-bound the complexity of the final structure. Figure 3.4 shows the pseudocode of the greedy algorithm described above. Figure 3.5 shows an example sequence of operators to learn the Bayesian network structure from Fig. 3.2.

At the beginning of this section we said that searching for the best Bayesian network is NP-complete. This leads to an interesting observation: BOA is a search algorithm that uses another search algorithm, BOA searches within a search. Why should the problem of learning the structure of Bayesian networks be simpler than the original optimization problem BOA attempts to solve? It is premature to discuss this topic in great detail, but theoretical results of Chap. 4 will show that even if the subproblems in a proper problem decomposition deceive GAs away from the optimum, the signal for constructing a Bayesian network will still lead in the right direction. The reason for this is that the search for a good model does not attempt to solve the optimization problem, but instead it learns interactions between the variables in a performance measure for the problem. Furthermore, BOA does not require the best network, it only needs a network that encodes all important interactions in a problem (or most of them). In addition to the these two arguments, an

```
Greedy algorithm for network construction
    initialize the network B (e.g., to an empty network);
    done := 0;

    repeat
        O = all simple graph operations applicable to B;
        if there exists an operation in O that improves score(B) {
            op = operation from O that improves score(B) the most;
            apply op to B;
        }
        else
            done := 1;

    until (done=1);

    return B;
```

Fig. 3.4. The pseudocode of the greedy algorithm for learning the structure of Bayesian networks

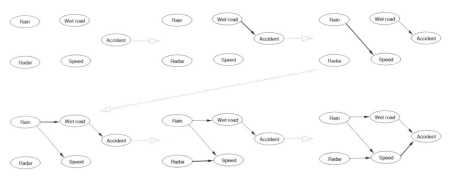

Fig. 3.5. An example sequence of steps leading to the network structure shown earlier in Fig. 3.2. For the sake of simplicity, only edge additions are used in this example

array of results presented throughout this book will support the above claim empirically.

3.4 Sampling Bayesian Networks

Once the structure and parameters of a Bayesian network have been learned, new candidate solutions are generated according to the distribution encoded by the learned network (see Equation (3.1)).

The sampling can be done using the probabilistic logic sampling of Bayesian networks [80], which proceeds in two steps. The first step computes an ancestral ordering of the nodes, where each node is preceded by its parents.

```
Algorithm for creating the ancestral ordering of the variables
  for each variable i {
    mark(i) := 0
  }
  k := 1;
  while (k <= number of variables) {
    Find variable i whose every parent x in B has mark(x)=1;
    ordered(k) := i;
    k := k+1;
    mark(i)=1;
  }
  return ordered;
```

Fig. 3.6. The pseudocode of the algorithm for computing the ancestral ordering of the variables in the Bayesian network

```
Probabilistic logic sampling of a Bayesian network B
  idx = ancestral_ordering(B);
  while (more instances needed) {
    for i=1 to length(idx) {
      generate variable idx(i) using p(idx(i)|parents(idx(i)));
    }
  }
```

Fig. 3.7. The algorithm for sampling a given Bayesian network starts by ordering the variables according to the dependencies, yielding an ancestral ordering. The variables of each new solution are generated according to the ancestral ordering using conditional probabilities encoded by the network

The basic idea is to generate the variables in such a sequence that the values of the parents of each variable are generated prior to the generation of the value of the variable itself. The pseudocode of the algorithm for computing an ancestral ordering of the variables is shown in Fig. 3.6.

In the second step, the values of all variables of a new candidate solution are generated according to the computed ordering. Since the algorithm generates the variables according to the ancestral ordering, when the algorithm attempts to generate the value of each variable, the parents of the variable must have already been generated. Given the values of the parents of a variable, the distribution of the values of the variable is given by the corresponding conditional probabilities. The second step is executed to generate each new candidate solution. A more detailed description of the algorithm for generating new candidate solutions is shown in Fig. 3.7.

3.5 Initial Experiments

This section presents the results of initial experiments using BOA on the aforementioned onemax and trap-5 problems. Additionally, BOA is tested on an additively separable deceptive function of order 3. All tested problems are of bounded difficulty – they can be decomposed into subproblems of bounded order. The performance of BOA is compared to that of the simple GA with uniform crossover and the stochastic hill climber using bit-flip mutation.

First, test problems are reviewed and discussed briefly. Next, experimental methodology is provided. Finally, the performance of BOA on the test problems is presented and compared to that of the simple GA with uniform crossover and the mutation-based hill climber.

3.5.1 Test Functions

Recall that onemax is defined as the sum of bits in the input binary string (see Equation (1.1)). The optimum of onemax is in the string of all 1s. Onemax is a unimodal function where the optimal value of each bit can be determined independently of other bits, and therefore it is easy to optimize. The purpose of testing BOA on onemax is to show that BOA can solve not only decomposable problems of bounded difficulty but also those problems that are simple and can be efficiently solved by the mutation-based hill climber.

Trap-5 is defined as the sum of 5-bit traps applied to non-overlapping 5-bit partitions of solution strings (see Sect. 1.5 on page 9 for a detailed definition). The partitioning is fixed, but there is no information about the positions in each partition revealed to BOA. Trap-5 cannot be decomposed into subproblems of order lower than 5 and therefore they can be used to analyze the scalability of BOA on nontrivial problems of bounded difficulty.

An additively separable deceptive function of order 3 [129] denoted by deceptive-3 is defined as the sum of single deceptive functions of order 3 applied to non-overlapping 3-bit partitions of solution strings. The partitioning is fixed, but there is no information about the positions in each partition revealed to BOA. The fitness contribution of each 3-bit partition is given by

$$
dec_3(u) = \begin{cases} 0.9 & \text{if } u = 0 \\ 0.8 & \text{if } u = 1 \\ 0 & \text{if } u = 2 \\ 1 & \text{otherwise} \end{cases} , \tag{3.6}
$$

where u is the number of ones in the input block of 3 bits. An n-bit deceptive-3 function has one global optimum in the string where all bits are equal to 1, and it has $2^{\frac{n}{3}} - 1$ local optima in strings where the bits corresponding to each deceptive partition are equal, but where the bits in at least one partition are equal to 0. Deceptive-3 represents yet another example of a nontrivial problem of bounded difficulty; however, since the order of subproblems in deceptive-3 is different compared to that in trap-5, a comparison of the performance of BOA on these two functions should reveal whether BOA performance depends on the order of subproblems in a proper problem decomposition or not.

3.5.2 Experimental Methodology

For all tested problems, 30 independent runs are performed and BOA is required to find the optimum in all the 30 runs. The performance of BOA is measured by the average number of fitness evaluations until the optimum is found. The population size is determined empirically by a bisection method so that the resulting population size is less than 10% from the minimum population size required to ensure that the algorithm converges in all the 30 runs. The bisection method used to determine the minimum population size is outlined in Fig. 3.8.

BIC is used to construct a Bayesian network in each generation, and the construction always starts with an empty network. Binary tournament selection without replacement is used in all the experiments. In most cases, better performance could be achieved by increasing selection pressure; however, the purpose of these experiments is not to show the best BOA performance, but to empirically analyze its scalability. The number of candidate solutions generated in each generation is equal to half the population size, and an elitist replacement scheme is used that replaces the worst half of the original population by offspring (newly generated solutions).

The GA with uniform crossover is also included in some of the results. All the parameters that overlap with BOA – except for population sizes – are set in the same fashion. Population sizes are also determined empirically by the bisection method. To maximize the mixing on onemax, the probability of applying crossover to each pair of parents is 1. On other problems, the crossover probability is 0.6. No mutation is used to focus on the effects of selectorecombinative search.

The performance of the mutation-based hill climber is also compared to that of BOA and the GA with uniform crossover. The mutation-based hill climber starts with a random solution. In each iteration, the hill climber applies bit-flip mutation to the current solution, and replaces the original solution by the new one if the new solution is better. The performance of the hill climber is computed according to the Markov-chain model of Mühlenbein [112], which provides the theory for computing both the optimal mutation rate as well as the expected performance.

3.5.3 BOA Performance

Figure 3.9 shows the number of fitness evaluations until BOA finds the optimum of onemax. The size of the problem ranges from $n = 100$ to $n = 500$ bits. The number of fitness evaluations can be approximated by $O(n \log n)$.

Bisection method for determining the minimum population size
// determine initial bounds

```
N := 100;
success := are 30 independent runs with pop. size N successful?
if (success) {
  while (success=1) {
    N := N/2;
    success := are 30 indep. runs with pop. size N successful?
  }
  low := N;
  high := 2*N;
}
else {
  while (success=0) {
    N := N*2;
    success := are 30 indep. runs with pop. size N successful?
  }
  low := N/2;
  high := N;
}

// use bisection to get within 10% of the minimum pop. size

while ((high-low)/low>=0.10) {
  N = (high+low)/2;
  success := are 30 independent runs with pop. size N successful?
  if (success=0)
    low := N;
  else
    high := N;
}

return high;
```

Fig. 3.8. Bisection method for determining minimum population size

Therefore, the results indicate that BOA can solve onemax in a near-linear number of evaluations.

Figure 3.10 shows the number of fitness evaluations until BOA finds the optimum of trap-5 and deceptive-3 functions. The size of trap-5 functions ranges from $n = 100$ to $n = 250$ bits, whereas the size of deceptive-3 functions ranges from $n = 60$ to $n = 240$ bits. In both cases, the number of fitness evaluations can be approximated by $O(n^{1.65})$. Therefore, the results indicate that BOA can solve trap-5 and deceptive-3 in a subquadratic number of evaluations. Furthermore, except for the trivial case with no interactions between problem variables in onemax, the order of subproblems in a proper problem

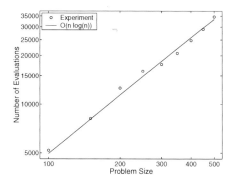

Fig. 3.9. The number of evaluations until BOA finds the optimum on onemax of varying problem size averaged over 30 independent runs. The number of evaluations can be approximated by $O(n \log n)$

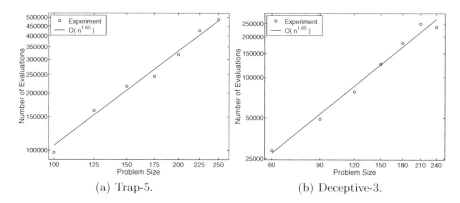

(a) Trap-5.   (b) Deceptive-3.

Fig. 3.10. The number of evaluations until BOA finds the optimum of trap-5 and deceptive-3 functions. The number of evaluations can be approximated by $O(n^{1.65})$ in both cases

decomposition does not seem to affect the order of growth of the number of fitness evaluations.

Experiments on other decomposable problems indicate similar performance. In the case of onemax, the performance of BOA is close to the expected performance with population-wise uniform crossover, which is the ideal crossover to use in this case. The performance gets slightly worse for other decomposable problems where the discovery of a proper decomposition is necessary. However, in all the cases, the number of evaluations until reaching the optimum appears to grow subquadratically with the number of variables in the problem (problem size).

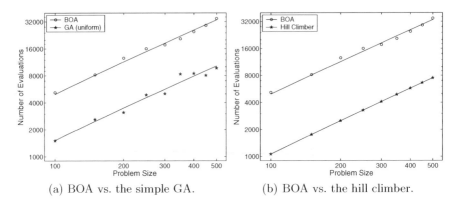

(a) BOA vs. the simple GA.   (b) BOA vs. the hill climber.

Fig. 3.11. The comparison of BOA, the simple GA with uniform crossover, and the mutation-based hill climber on onemax

3.5.4 BOA vs. GA and Hill Climber

For onemax, both the hill climber and the simple GA with uniform crossover can be expected to converge in $O(n \log n)$ evaluations. The performance of the simple GA can be estimated by the gambler's ruin population-sizing model [70] and onemax convergence model [118]. For the simple GA with one-point or n-point crossover, the performance would get slightly worse because of a slower mixing, which results in an increased number of generations. The performance of the hill climber can be estimated using the theory of Mühlenbein [112].

Figure 3.11 compares the performance of the simple GA and the mutation-based hill climber with that of BOA for onemax. In all cases, the total number of evaluations is bounded by $O(n \log n)$; however, the performance of BOA is about 3.57-times worse than the performance of the simple GA, and the performance of the simple GA is about 1.3-times worse than the performance of the hill climber.

The reason for the worse performance of BOA is that onemax can be solved by processing each bit independently, but BOA introduces unnecessary dependencies, which lead to increased population-sizing requirements according to the gambler's ruin population-sizing model [70]. The comparison of the performances of the simple GA and the hill climber is inconclusive, because choosing a different selection pressure would change the results of the simple GA. However, it is important to note that the number of evaluations for all the algorithms grows as $O(n \ln n)$. That means that even for those simple problems that are ideal for mutation and uniform crossover, the use of sophisticated search operators of BOA does not lead to a qualitative decrease in the performance.

Figure 3.12 compares the performance of the simple GA and the mutation-based hill climber with that of BOA for trap-5. The performance of the GA and the hill climber dramatically changes compared to onemax.

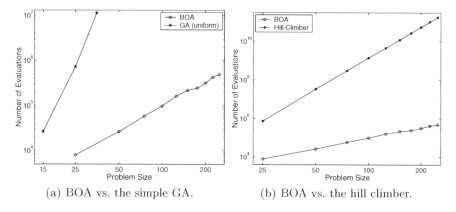

(a) BOA vs. the simple GA.   (b) BOA vs. the hill climber.

Fig. 3.12. The comparison of BOA, the simple GA with uniform crossover, and the mutation-based hill climber on trap-5

To solve trap-5, the simple GA requires exponentially large populations, because mixing is ineffective and requires exponentially large populations for innovative success [172]. Similar performance can be expected with other crossover methods, because the partitions of a proper problem decomposition are not located tightly in solution strings. Already for the problem of size 50, even population sizes of a half million do not result in reliable convergence; the figure only shows the results on problems of sizes 15, 20, and 25 bits. Therefore, the performance of the simple GA changes from $O(n \log n)$ for onemax to $O(a^n)$ where $a > 1$ for trap-5.

The performance of the hill climber changes from $O(n \log n)$ for onemax to $O(n^5 \ln n)$ for trap-5 [112]. In the general case, the complexity of the hill climber grows with the largest minimal order k of statistics required to solve the problem as $O(n^k \ln n)$.

Similar results can be observed for deceptive-3 (see Fig. 3.13). The number of evaluations for the simple GA grows exponentially, and the number of evaluations for the hill climber grows as $O(n^3 \log n)$.

3.5.5 Discussion

To summarize, there are three important observations:

(1) **All on onemax.** BOA, GA, and the hill climber find the optimum of onemax in approximately $O(n \log n)$ evaluations. However, BOA is outperformed by the simple GA and the hill climber within a constant factor.

(2) **GA and the hill climber on trap and deceptive functions.** The performance of the simple GA and the mutation-based hill climber significantly suffers from an increased order of an adequate problem decomposition. Both algorithms become intractable for problems of moderate difficulty.

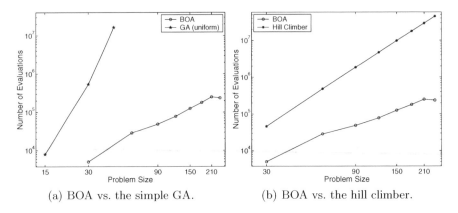

(a) BOA vs. the simple GA.   (b) BOA vs. the hill climber.

Fig. 3.13. The comparison of BOA, the simple GA with uniform crossover, and the mutation-based hill climber on deceptive-3 functions

(3) **BOA.** BOA requires a subquadratic number of evaluations until it finds the optimum of all tested problems of bounded difficulty; these results are supported by theory in the next chapter. If identifying variable interactions is unnecessary, $O(n \log n)$ evaluations until convergence can be expected. If identifying variable interactions is necessary, the performance is subquadratic. Furthermore, the performance of BOA does not depend on the location of positions corresponding to each subproblem in a proper decomposition.

4

Scalability Analysis

The empirical results of the last chapter were tantalizing. Easy and hard problems were automatically solved without user intervention in polynomial time. This raises an important question: How is BOA going to perform on other problems of bounded difficulty?

The purpose of this chapter is to analyze the scalability of BOA on problems of bounded difficulty. As a measure of computational complexity, the chapter considers the number of evaluations of the objective function until the optimum is found with high confidence. The total number of evaluations is computed by (1) estimating an adequate population size that ensures reliable convergence to the optimum, (2) approximating the number of generations until convergence, and (3) making a product of these two quantities. Empirical results are then compared to the developed theory.

The chapter starts by arguing that the number of evaluations is a reasonable measure of computational complexity of BOA. Section 4.2 provides background of the GA population-sizing theory. Section 4.3 focuses on the BOA population-sizing theory. Section 4.4 discusses background of GA time-to-convergence theory, which estimates the number of generations until convergence assuming a sufficiently large population. Section 4.5 presents the convergence model for population-wise uniform crossover (univariate marginal distribution) on onemax and discusses how the model can be extended to the more general case of BOA. Two bounding types of scaling the subproblems in a problem decomposition are considered: (1) uniform scaling, and (2) exponential scaling. The scaling in any decomposable problem of bounded difficulty is expected to lie between the two cases. Finally, Sect. 4.6 puts all the pieces of the scalability puzzle together and validates the developed theory with experiments.

Martin Pelikan: *Hierarchical Bayesian Optimization Algorithm*, StudFuzz **170**, 49–87 (2005)
www.springerlink.com © Springer-Verlag Berlin Heidelberg 2005

4.1 Time Complexity and the Number of Evaluations

One of the important characteristics of black-box optimization algorithms is the number of evaluations until reliable convergence to the optimum. There are two primary reasons why considering the number of evaluations makes sense. The first reason is that in complex real-world applications, evaluating a candidate solution is often computationally expensive and, as a result, the time spent in evaluation usually overshadows the time spent in the remaining parts of the optimization method. The second reason is that per-evaluation computational overhead grows usually as a low-order polynomial of the problem size. Good scalability with respect to the number of evaluations thus implies good scalability with respect to standard measures of computational complexity.

Let us assume that in each generation the number of new candidate solutions that must be evaluated is equal to a certain proportion of the size of the original population before selection,

$$E_g = cN \, , \tag{4.1}$$

where $c \in (0, 1]$ is a constant and N is the size of the population before selection. For example, if the size of the offspring population is equal to the size of the original population before selection, then $c = 1$. On the other hand, if only half of the original population is replaced by offspring, then $c = 0.5$.

The number of evaluations E until convergence is then given by

$$E = N + E_g G = N + c(N \times G) \, , \tag{4.2}$$

where G is the number of generations until convergence. Note that N, G, and E are functions of the problem size and the type of the problem, and that the focus is on the growth of E with respect to the problem size. Since the number of generations does not *decrease* with the problem size and other problem parameters,

$$E = O(N \times G) \, . \tag{4.3}$$

Therefore, to compute the growth of E with respect to the problem size, it is sufficient to compute the required population size and the number of generations until convergence, both expressed in terms of the problem size.

The following section provides background of the GA population-sizing theory. Section 4.3 presents the BOA population-sizing model. Section 4.4 provides background of GA theory for estimating the number of generations until convergence. Section 4.5 focuses on the number of generations until reliable convergence in BOA. Finally, Sect. 4.6 combines the pieces of theory to compute the overall number of evaluations required by BOA on decomposable problems of bounded difficulty.

4.2 Background of GA Population-Sizing Theory

GA population-sizing theory attempts to estimate an adequate population size for ensuring reliable convergence to the optimum. There are three important factors influencing the population sizing in GAs (assuming that crossover combines candidate solutions effectively):

1. **Initial supply.** The population must be large enough to ensure that there is a sufficient supply of alternative solutions to each subproblem in an adequate problem decomposition.
2. **Decision making.** The population must be large enough to ensure that decision making between alternative solutions to each subproblem is not misled by the noise from the remaining subproblems and that the best partial solution (the building block or BB) indeed wins.
3. **Genetic drift.** The population must be large enough to ensure that if the subproblems converge in several phases, there is an adequate initial supply of alternative solutions to each subproblem once the algorithm gets to the beginning of the subproblem's phase.

The next section discusses population-sizing models that focus on the initial supply of building blocks (BBs). Subsequently, population-sizing models concerned with the decision making are reviewed briefly. Finally, models that deal with genetic drift are examined.

4.2.1 Having an Adequate Initial Supply of BBs

Recall the GA simulation for onemax presented in Sect. 1.3. Without making sure that there are enough 1s in each position of the initial population, the GA could not find the optimum (regardless of how well or how badly it would recombine). To ensure that the solution can be found by combining bits and pieces of promising solutions, it is necessary that there is enough raw material to start with. Initial-supply population-sizing models [57, 61, 83] focus on the initial supply of this raw material and bound the population size so that the initial population contains enough BBs to enable the selectorecombinative search of GAs to juxtapose these BBs and "build" the optimum.

Let us start with some simple mathematics to motivate initial-supply models, followed by a short review of past work on this topic. Assuming that the initial population of N binary strings is generated at random, the expected number of copies of any partial solution of order k is

$$m(BB_k) = \frac{N}{2^k} . \tag{4.4}$$

This suggests that to ensure a fixed number of copies of each BB, the population size should grow at least exponentially with the BB size. Of course, the actual number of copies can deviate from its expected value, because the initial population is generated at random; this fact must be considered to ensure that the initial supply is ensured with high confidence.

The importance of the initial supply of BBs was first recognized by Holland [83], who computed the number of BBs that receive a specified number of copies in the initial population using Poisson distribution. Goldberg [54] refined Holland's model by using binomial distribution and applied the resulting model to population sizing. Reeves [148] proposed an initial-supply population-sizing model for BBs of unit size and fixed cardinality. Poli et al. [143] looked at the required number of copies of each BB to prevent its loss. Most recently, Goldberg et al. [61] computed the necessary population size to ensure that every BB gets at least one copy in the initial population using alphabets of arbitrary cardinality, yielding a lower bound on the population size proportional to $2^k(k + \log n)$.

4.2.2 Deciding Well Between BBs and Their Competitors

Besides having an adequate initial supply of BBs, another important factor that determines the success of GAs is the one of *deciding well* between BBs and their competitors. Naturally, each BB should get more copies in the offspring population than its competitors do. However, as it was recognized by Holland [82], the decision making in GAs has statistical nature, and the population size must be set accordingly to ensure that good decisions are made with high probability. Holland illustrated the statistical nature of decision making using a 2^k-armed bandit model. Later, De Jong [34] developed equations for the 2-armed bandit and noted the importance of *noise* in the GA decision making. In short, the decision-making for a particular BB is affected by *noise* from the fitness contributions of the remaining partitions of the problem decomposition (the context); this noise is often referred to as *collateral noise*.

The effects of collateral noise can be illustrated by looking at the GA simulation on onemax shown in Fig. 1.2. Although the last bit of 00001 is contained in the optimum and it increases the fitness of each solution independently of its context, the fitness of 00001 is worse than that of 10010. If 00001 and 10010 compete in a tournament, the proportion of 1s in the last position will *decrease*. However, on average, the performance of solutions with a 1 in the last position should increase. Decision-making population-sizing models focus on computing an adequate population size to ensure that good decisions are made; in the onemax case, good decision making should increase the proportion of 1s in each position of the population over time.

Goldberg and Rudnick [60] computed the variance of collateral noise using Walsh analysis and considered the ramifications of the result for the decision making. The variance of collateral noise formed the basis of the first practical decision-making population-sizing model [58]. The proposed model reduced decision making to the two best partial solutions of a subproblem – the BB and its toughest competitor (second best partial solution in the same partition). It estimated the required population size so that each BB wins over its best competitor in the same partition; once that is ensured, it is natural

to expect the BB to win over its remaining competitors in the same partition as well. This model was a bit pessimistic, it required each BB to win in
the first generation. Assuming that the problem is decomposable into similar
subproblems of bounded order, the resulting population-sizing estimate was
computed as

$$N = 2c(\alpha)2^k m' \frac{\sigma_{bb}^2}{d^2} , \tag{4.5}$$

where $c(\alpha)$ is the square of the ordinate of a unit normal distribution where
the probability equals to α; α is the probability of failure; k is the order of the
considered BB; m' is one less than the number m of BBs (i.e. $m' = m - 1$); σ_{bb}^2
is the root mean square (RMS) fitness variance of the fitness contributions in
the BB's partition; and d is the difference between the fitness contributions
of the BB and its toughest competitor.

Harik et al. [71, 74] refined the above population-sizing estimate by elim-
inating the requirement for successful decision making in the first generation, and modeling subsequent generations using gambler's ruin model in
one dimension [42]. Assuming perfect mixing (e.g., population-wise uniform
crossover on onemax) the bound on the population size sufficient to find a
solution containing each BB with the probability $(1 - \alpha)$ was reduced to

$$N = -2^{k-1} \ln(\alpha) \frac{\sigma_{bb}\sqrt{\pi m'}}{d} . \tag{4.6}$$

Empirical results with tightly-encoded deceptive BBs and the two-point
crossover matched the theory very well [74]. Thus, with perfect mixing, the
required population size in GAs grows proportionally to the square root of
the number of BBs in a problem. The gambler's ruin population-sizing model
was later extended to accommodate noise in the fitness function [71].

An empirical population-sizing model for onemax, truncation selection,
and uniform crossover was discussed by Mühlenbein and Schlierkamp-Voosen
[118]; this model agrees with the Gambler's ruin model for this special case
and estimates the population size as $O(\sqrt{n}\log n)$.

4.2.3 Genetic Drift

Genetic drift causes some partial solutions to be lost due to randomness of
selection even if their fitness values are the same or better than those of
their competitors. This could become a problem if the contributions of the
different subproblems are scaled so that the subproblems converge in several
phases, where in each phase only some subproblems matter. Consequently,
when the subproblem's phase starts, some partial solutions corresponding to
this subproblem may have already been eliminated. Much of the following
discussion on the drift population-sizing models is motivated by the works
of Rudnick [156], Thierens et al. [177], Lobo et al. [101], Rothlauf [152, 153],
and Albert [2].

To illustrate the importance of genetic drift in sizing the populations for some problems, it is helpful to consider the *binary integer* fitness function [156] defined as

$$f_{bin}(X) = \sum_{i=1}^{n} 2^{n-i} X_i \ , \tag{4.7}$$

where $X = (X_1, \ldots, X_n)$ is the input binary string of n bits. The binary integer fitness decodes the binary number and uses the decoded value to determine its fitness. For example, $f_{bin}(001) = 0 \times 2^2 + 0 \times 2^1 + 1 \times 2^0 = 1$ and $f_{bin}(110) = 1 \times 2^2 + 1 \times 2^1 + 0 \times 2^0 = 6$.

Similarly as in onemax (see Equation (1.1) on page 5) the optimum of the binary integer is in the string of all ones and each bit can be considered independently of its context. However, in the binary integer, each bit contributes to the fitness more than all the remaining bits to its right together. Consequently, selection will always put pressure only on one or a very small subset of bits and the GA will converge sequentially, one or a few bits at the time. This can be illustrated on a random population of 4 strings of 8 bits, which may look as follows:

```
01101110
00001101
10010100
11011100
```

In the above population, comparing any pair of solutions can be done using only the first two bits; the remaining 6 bits can be ignored. Once the first two bits of solution strings converge or nearly converge, the subsequent bits become relevant. Since each bit becomes relevant only if there are two solutions that match each other in all bits to the left of the considered bit, it is easy to see that the number of bits that are relevant for selection from a random population grows logarithmically with the population size.

Consequently, at any point in time, some bits are already fixed, one or a couple of bits are just in the process of converging, and the remaining bits fluctuate randomly due to the stochasticity of selection [156]. The random fluctuations of the partial solutions whose fitness contributions are too low are often called *genetic drift*. See Fig. 4.1 to visualize this process.

Due to genetic drift, once the GA starts optimizing the last bit (which did not matter in the selection process so far) its optimal value (in this case, 1) might be already gone. The drift population-sizing models ensure that when a particular subproblem comes into play, the population still contains enough partial solutions to this subproblem to find the global optimum. There are two bounding cases of scaling the subproblems: (1) All the subproblems are scaled the same and converge at the same time, and (2) there is one phase per variable, and the variables converge sequentially. Problems from the first case are taken care of by the initial-supply and decision-making models. The problems between the two extremes are discussed next.

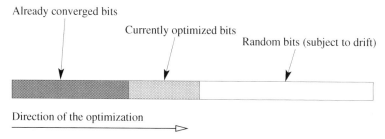

Fig. 4.1. On problems with exponentially scaled subproblems, the variables converge sequentially. At any point in time, only a small subset of variables affects selection; other variables have either converged already or their influence on the fitness is too small to affect selection

It can be shown [4, 62, 101] that, assuming that if the considered subproblem does not play role in selection at the time, the expected time for losing the partial solution due to genetic drift (the drift time) is given by

$$t_{drift} = cN \, , \tag{4.8}$$

where N is the population size, and the constant c depends on the initial proportion of the considered partial solution. Note that by enlarging the population size, any drift time can be achieved. In other words, the time until any partial solution is eliminated due to genetic drift can be increased arbitrarily by enlarging the population.

To find out how long the waiting must be, it is necessary to compute the number of generations until the start of each phase. The number of phases can be upper-bounded by the size of the problem (since in each phase at least one variable must fully converge):

$$n_{phase} \leq n \, . \tag{4.9}$$

Let us denote the number of subproblems of bounded order that play role in the ith phase by n_i, where $i \in \{1, \ldots, n_{phase}\}$. Each variable must be contained in at most one phase:

$$\sum_i n_i \leq n \, . \tag{4.10}$$

The number of generations required to converge in the ith phase can be bounded by [118, 175]

$$t_i = O(\sqrt{n_i}) \, . \tag{4.11}$$

Therefore, the number of generations until the ith phase starts is given by

$$T_i = \sum_{k=1}^{i-1} t_i = \sum_{k=1}^{i-1} a_i \sqrt{n_i} \, , \tag{4.12}$$

where a_i are constants depending on the type of the problem and the selection method used. We must now ensure that no partial solution in the ith problem drifts earlier than T_i for every phase i:

$$t_{drift} > T_i .$$ (4.13)

Therefore, we require that for all phases i, the following equation must be satisfied

$$cN \geq \sum_{k=1}^{i-1} a_i \sqrt{n_i} .$$ (4.14)

The right-hand side of the last equation can grow at most linearly with the size of the problem, because

$$\sum_{k=1}^{i-1} a_i \sqrt{n_i} \leq \sum_{k=1}^{i-1} a_i n_i = O(n) .$$ (4.15)

Therefore, $cN \geq dn$, where c and d are constants, yielding

$$N = O(n) .$$ (4.16)

To summarize, the bound on the population size from the above equations can range from $O(\sqrt{n})$ (when all the subproblems are uniformly scaled as in onemax) to $O(n)$ (when each subproblem corresponds to a special phase as in the binary integer). The population size should thus grow at most linearly with the problem size.

The following section focuses on the population sizing in BOA.

4.3 Population Sizing in BOA

There are four factors that influence the population sizing in BOA [131]. The first three factors were discussed in the background part of this chapter; these factors can be adopted from the GA population-sizing theory [54, 57, 58, 61, 74, 83]. However, the GA theory assumes that crossover combines and preserves the building blocks properly; in the terminology of BOA, the probabilistic model is assumed to encode an adequate problem decomposition. In addition to the first three factors, BOA must be capable of learning a proper, non-misleading, problem decomposition. Once this is ensured, the results of the GA population-sizing theory considering the first three factors can be applied.

This section presents the theory for sizing the populations in BOA so that a good model is found [131, 140, 141]. To enable the development of the population sizing model, the theory makes a number of simplifying assumptions. One of these assumptions is that the theory requires that, with high confidence, a good model is expected to be found in the first generation. Despite the simplifying assumptions, the resulting bound on the population size fits the actual empirical results well.

4.3.1 Road Map to BOA Population-Sizing Model

The section starts by discussing sources of statistical dependencies in the population of promising solutions. Section 4.3.3 relates collateral noise to the problem size, states the important assumptions of the BOA population-sizing model, and provides basic notational conventions used in the BOA population-sizing theory. Section 4.3.4 discusses the problem of deciding between adding and not adding an edge into the Bayesian network and defines the *critical population size*, which is the minimum population size for finding the dependency under consideration. Section 4.3.5 computes the probabilities of one partition of the problem decomposition after applying binary tournament selection. The section shows that any two positions can be treated separately even if they are contained in a bigger partition.

Section 4.3.6 computes the critical population size for the general two-bit case. Section 4.3.8 discusses the effect of using finite populations on the difference between the expected frequencies and the actual ones after applying binary tournament selection. Section 4.3.9 compares the developed theory to the empirical results. Section 4.3.10 summarizes and discusses the results of the BOA population-sizing theory. The chapter then continues by discussing the number of generations until BOA convergence.

If the reader is not interested in the detailed theory, he or she may skip over most of this section without losing track and continue with the section summary on page 78.

4.3.2 Finding a Proper Model: The Good, the Bad, and the Ugly

Consider two string positions, X_1 and X_2. If the fitness contributions of X_1 and X_2 are independent as in onemax, the model should not contain an edge between the two variables. On the other hand, if the contributions of X_1 and X_2 do depend on each other, it might be necessary to consider the dependency between the two variables. Not all nonlinearities must be covered; however, some nonlinearities could lead to deception like in trap-5 described in Chap. 1. To avoid deception, BOA should consider as many nonlinearities as possible.

However, the construction of a model is guided by statistical dependencies and independencies over subsets of variables, while nonlinearities that must be detected are in the fitness. To find a model that encodes nonlinearities in the problem, it is necessary to ensure that selection "transforms" nonlinearities in the fitness into statistical dependencies in the population of promising solutions.

If the fitness contributions of X_1 and X_2 depend on each other, it can be expected that selection will lead to a detectable statistical dependency between the two variables. However, contrary to intuition, all commonly used selection methods also introduce the statistical dependencies between the variables whose fitness contributions are independent. This happens even if an infinite population is used. Let us illustrate this on an example. Consider

Table 4.1. The proportions of the solutions on a 2-bit onemax before and after binary tournament selection. The population is assumed to be infinite

(a) Initial (random) population		(b) Population after binary tournament	
X	$p(X)$	X	$p(X)$
00	0.25	00	0.0625
01	0.25	01	0.25
10	0.25	10	0.25
11	0.25	11	0.4375

a 2-bit onemax and an infinite population. As shown in Table 4.1(a), all the solutions 00, 01, 10, and 11 will occupy 25% of the initial population. After performing binary tournament selection, the frequencies will change as shown in Table 4.1(b). If the two positions were statistically independent, the following equation would have to be satisfied:

$$p(11) = p(1*)p(*1) ,$$

where $p(1*)$ and $p(*1)$ denote the total probability of a 1 on the first and second positions, respectively. Substituting the probabilities from Table 4.1 yields $p(11) = 0.4375$ on the left-hand side of the above equation, and $p(1*)p(*1) = 0.6875^2 \approx 0.4727$ on the right-hand side. Clearly, $0.4375 \neq 0.4727$ and therefore the two variables are *not* statistically independent, even though their fitness contributions *are* independent. A similar dependency will be created by practically all common selection methods except for Boltzmann selection.

Therefore, there are two kinds of statistical dependencies after selection; some dependencies are introduced by the nonlinearities in the fitness, while some dependencies are introduced by selection only. It is important that BOA discovers those dependencies that correspond to the nonlinearities (good dependencies), and that it ignores those dependencies that are introduced by selection only (bad dependencies). In particular, it is desirable that the following two conditions are satisfied:

1. The statistical dependencies corresponding to the fitness nonlinearities are significantly stronger than the statistical dependencies introduced by selection only.
2. The stronger the nonlinearity, the stronger the statistical dependency.

This section shows that both the conditions are satisfied. To make the computation tractable, the section first assumes that the frequencies are equal to their expected values with an infinite population, although the focus is on the sufficient population size to discover the good dependencies and avoid the bad ones. Next, the assumption about the accuracy of the frequencies is justified by computing the minimal bound on the population size that ensures that with high confidence, the frequencies will be sufficiently close to those expected with an infinite population.

4.3.3 Assumptions and Notation

To make the theoretical analysis tractable, we make several assumptions about the problem. First, we assume that the fitness function is defined as the sum of the subfunctions applied to disjoint subsets of the variables of order k and that all the subfunctions are the same. In some cases, it is possible to apply the results of the theory also to the case where the subfunctions overlap; however, in most cases the theory would have to be extended to incorporate the effects of the overlap.

A particular partition of the problem decomposition is considered. Without loss of generality, we denote the variables in the considered partition by $X = (X_1, \ldots, X_k)$ or $Y = (Y_1, \ldots, Y_k)$ and their instantiations (blocks of k bits or partial solutions) by $x = (x_1, \ldots, x_k)$ and $y = (y_1, \ldots, y_k)$. The fitness contribution of X and Y is denoted by $g(X_1, \ldots, X_k) = g(X)$ and $g(Y_1, \ldots, Y_k) = g(Y)$, respectively. We denote the total fitness of the solutions containing the block $x = (x_1, \ldots, x_k)$ by $F(x)$ (note that $F(x)$ is a random variable) and we assume that the contributions of the remaining variables can be modeled by a normal distribution with the variance proportional to the size of the problem:

$$F(x) \sim N\left(\mu_x, \sigma_N^2\right) , \tag{4.17}$$

where μ_x is the average fitness of the solutions containing $x = (x_1, \ldots, x_k)$, and σ_N^2 is the variance of the fitness contributions of the remaining variables (collateral noise). The normal distribution assumption is justified because recombination in GAs and PMBGAs has normalizing effect [118]. Additionally,

$$\sigma_N^2 \propto n , \tag{4.18}$$

where n is the size of the problem. The above assumption can be justified by the central limit theorem for all decomposable problems of bounded difficulty where the fitness contribution of each subproblem is of the same magnitude and the order of the subproblems is bounded by a constant. It is straightforward to verify that Equation (4.18) holds for onemax and trap-5. If the magnitude of the contributions varies from one subproblem to another, the population size required for building a good model should decrease, because in each generation only a subset of the subproblems will matter. Although the subproblems do not have to be the same for different problem sizes, it is assumed that both the strength of the nonlinearities in the subproblems as well as the signal for each BB defined as the difference between the fitness contributions of the BB and its toughest competitor, are lower bounded by a constant.

By $p(x) = p(x_1, \ldots, x_k)$, we denote the probability of the partial solution x in the selected population of promising solutions. The probability is computed as the relative frequency of the partial solution x in the selected population. The probability of other partial solutions is denoted in a similar fashion. For example, $p(x_1, x_2)$ denotes the probability of the solutions with $X_1 = x_1$ and $X_2 = x_2$, $p(x_2)$ denotes the probability of the solutions with $X_2 = x_2$, and so forth. The probability distribution of $p(x)$ is denoted by $p(X)$. The probabilities *before* selection are denoted by $p_{init}(x)$.

Additionally, as mentioned above, in many derivations we assume that the observed probabilities are equal to their expected values with an infinite population, although in practice we can expect additional noise due to the finite size of the population. The assumption is later justified by bounding the population size so that the frequencies are close enough to their expected values with high confidence.

We consider only binary tournament selection (see page 3). Although the results seem to hold with other selection methods, the theoretical analysis becomes intractable. The empirical results for tournament selection with bigger tournaments are presented to justify this claim. The selected population is assumed to be of the same size as the population before selection. BIC metric is used to evaluate the model in most of the analysis.

External noise in the fitness function can be incorporated into the theory in a straightforward manner, if the noise can be approximated by a zero-mean normal distribution. If the variance of external noise does not grow faster than linearly with the size of the problem, all the above assumptions will remain satisfied. However, if external noise grows faster than linearly with the size of the problem, the theory would have to be modified slightly; however, all the modifications are straightforward and are therefore omitted for the sake of simplicity.

4.3.4 Edge Additions and the Critical Population Size

Consider the decision making between the following two cases:

(1) Add the edge from X_2 to X_1.
(2) Do not add the edge from X_2 to X_1.

To decide whether to add or not add the edge, we must compare the values of the used scoring metric for the current network with and without the edge, and choose the better alternative. Since both MDL and Bayesian metrics are decomposable, it is sufficient to look at the term corresponding to the node X_1.

There are two cases. Figure 4.2 illustrates both cases. In the first case (see part (a) in the figure), X_1 is isolated; in the second case (see part (b) in the figure), several edges that end in X_1 already exist in the network. The first case is first analyzed in detail. Section 4.3.7 extends the results of the analysis to the second, more general, case.

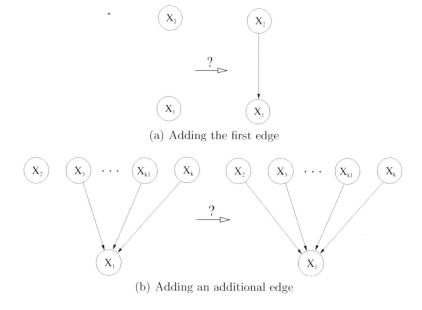

(a) Adding the first edge

(b) Adding an additional edge

Fig. 4.2. There are two cases to consider when making a decision between adding and not adding an edge. The first case assumes that there are no other edges into the terminal node of the considered edge, the second assumes that there are a number of such edges

The score assigned by BIC to X_1 without the edge from X_2 to X_1 is given by

$$BIC(X_1) = -H(X_1)N - \frac{\log_2 N}{2} , \qquad (4.19)$$

where $H(X_1)$ is the entropy of X_1, and N is the number of selected solutions (the size of the selected population). After adding an edge from X_2 to X_1, the new score for X_1 is given by

$$BIC(X_1 \leftarrow X_2) = -H(X_1|X_2)N - \log_2 N , \qquad (4.20)$$

where $H(X_1|X_2)$ is the conditional entropy of X_1 given X_2. For an addition of the edge $X_2 \rightarrow X_1$, the following inequality must be satisfied:

$$BIC(X_1 \leftarrow X_2) > BIC(X_1) . \qquad (4.21)$$

By substituting the equations (4.19) and (4.20) into the last equation, we get

$$(H(X_1) - H(X_1|X_2)) N - \frac{\log_2 N}{2} > 0 . \qquad (4.22)$$

Let us denote the difference between the marginal and conditional entropies of X_1 by D:

$$D = H(X_1) - H(X_1|X_2) \, . \qquad (4.23)$$

If X_1 and X_2 are not statistically independent, D is strictly positive. The positivity of D is later supported by an exact calculation of D in the general case; in particular, D is shown to be positive if X_1 and X_2 contribute to the fitness in some way. If X_1 and X_2 do not influence the fitness, $D = 0$ and the model will not add any dependency between the two variables.

Since $D > 0$ and the linear term in Equation (4.22) grows faster than the logarithmic one, Equation (4.22) will be satisfied for a large enough N. Intuitively, when the two variables are not independent, for a big enough population size, the dependency should be discovered. We call the sufficient population size for the discovery of the dependency $X_2 \to X_1$ the *critical population size* and denote it by N_{crit}. To determine N_{crit}, the following equation must be solved for N:

$$N - \frac{\log_2 N}{2D} = 0 \, . \qquad (4.24)$$

The above equation has two solutions but there is no closed form for either of these solutions. The dependency is discovered for the population sizes lower than the first (lower) solution or greater than the second (greater) one. The first solution is approximately equal to $1 + 2D \ln 2$. However, since even for small problems the value of D is very small, this solution is of no interest in our case. N_{crit} is therefore defined as the larger of the two solutions of the last equation.

If the ratio $\frac{1}{2D}$ is large enough, the larger of the two solutions of the above equation follows a power law is thus of the form $\alpha \left(\frac{1}{2D}\right)^{\beta}$, where $\alpha \sim 8.34$ and $\beta \sim 1.05$. More specifically,

$$N_{crit} = 8.33 \left(\frac{1}{2D}\right)^{1.05} = 4.027 \, D^{-1.05} \qquad (4.25)$$

Since for an increasing problem size the magnitude of D decreases inversely proportionally to the number of decision variables in the problem or even faster, $\frac{1}{2D}$ is going to be large and the above approximation can be used to determine N_{crit}. Figure 4.3 shows the numerical solution and its approximation using Equation (4.25). The solutions corresponding to the two methods are practically indistinguishable.

In order to apply our results to the scale-up behavior of BOA with BIC metric, we are interested in the *growth* of N_{crit} with respect to the size of the problem (the number of decision variables). Using the approximation given in Equation (4.25), it can be shown that the growth of N_{crit} is proportional to the growth of $\frac{1}{2D}$ and therefore to determine the growth of N_{crit} with respect to the size of the problem, it is sufficient to compute the growth of $\frac{1}{2D}$ with respect to the same parameter.

The following section starts by computing the probabilities of the partial solutions after applying binary tournament selection. These probabilities are

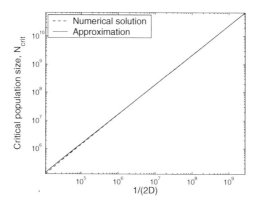

Fig. 4.3. Critical population size with respect to the ratio $\frac{1}{2D}$ for BIC metric

then used to determine the growth of D in two cases: (1) the fitness contributions of X_1 and X_2 are correlated, and (2) the fitness contributions of X_1 and X_2 are independent. The two cases are distinguished by a parametrization of the fitness contribution of X_1 and X_2. The growth of D is then substituted to Equation (4.25) to determine the growth of N_{crit}.

4.3.5 Block Probabilities After Binary Tournament

The initial population is generated at random with the uniform distribution and therefore the probability of any instantiation of the variables in the considered block of k binary variables is given by

$$p_{init}(x) = \frac{1}{2^k} , \qquad (4.26)$$

where $x = (x_1, \ldots, x_k)$ denotes the bits in the considered block.

Binary tournament selection selects two parents at random and chooses the one with the better fitness. Denote the probability of a tournament between the competing blocks x and y in a particular order by p_{tourn} (note that x and y correspond to the same partition in the problem). Using Equation (4.26),

$$p_{tourn} = p_{init}(X)p_{init}(Y) = \frac{1}{2^{2k}} . \qquad (4.27)$$

The ordering of x and y in the tournament does not affect the result of the tournament and, therefore, the probability of x after binary tournament selection is given by

$$p(x) = \sum_{y_1, \ldots, y_k} 2p_{tourn}p(F(x) > F(y)) , \qquad (4.28)$$

where $F(x)$ and $F(y)$ denote the fitness distribution of the blocks x and y, respectively (see Equation (4.17)); and $p(F(x) > F(y))$ denotes the probability that the solution with x wins the tournament over the solution with y. The probability of x winning the tournament over y can be rewritten as

$$p(F(x) > F(y)) = p(F(x) - F(y) > 0) . \tag{4.29}$$

Since both $F(x)$ and $F(y)$ are normally distributed (see Equation (4.17)), $F(x) - F(y)$ follows the normal distribution with the mean equal to the difference of the individual means of $F(x)$ and $F(y)$, and the variance equal to the sum of the variances of the two distributions. The difference of the mean fitness of $F(x)$ and $F(y)$ is equal to the difference of their contributions to the overall fitness denoted by $g(x)$ and $g(y)$, because the $F(x)$ and $F(y)$ are not correlated (due to the assumptions) and, consequently, the contributions of the remaining bits cancel out. Thus,

$$F(x) - F(y) \sim N(g(x) - g(y), 2\sigma_N^2) . \tag{4.30}$$

That yields

$$p(F(x) > F(y)) = \Phi\left(\frac{g(x) - g(y)}{\sqrt{2}\sigma_N}\right) , \tag{4.31}$$

where $\Phi(x)$ denotes the cumulative probability density function of the zero-mean normal distribution with the standard deviation of 1. The resulting probability of x after binary tournament selection is thus given by

$$p(x) = \sum_y \frac{1}{2^{2k-1}} \Phi\left(\frac{g(x) - g(y)}{\sqrt{2}\sigma_N}\right) . \tag{4.32}$$

The ratio of the fitness differences to the deviation of collateral noise decreases with the problem size and is usually a very small number for moderate-to-large problems. Therefore, a linear approximation of the cumulative density function in the above equation can be used, where $\Phi(x) = \frac{1}{2} + \frac{x}{\sqrt{2\pi}}$, yielding

$$p(x) = \sum_{y_1,\ldots,y_k} \frac{1}{2^{2k-1}} \left(\frac{1}{2} + \frac{g(x) - g(y)}{2\sqrt{\pi}\sigma_N}\right) . \tag{4.33}$$

In most cases we are interested only in the marginal probabilities of the instances x_1 and x_2 of the first two variables. These can be computed by marginalization:

$$\begin{aligned}
p(x_1, x_2) &= \sum_{x_3,\ldots,x_k} p(x) \\
&= \sum_{x_3,\ldots,x_k} \sum_{y_1,\ldots,y_k} \frac{1}{2^{2k-1}} \left(\frac{1}{2} + \frac{g(x) - g(y)}{2\sqrt{\pi}\sigma_N}\right) \\
&= \frac{1}{2^{2k-1}} \left(2^k \sum_{x_3,\ldots,x_k} \frac{g(x)}{2\sqrt{\pi}\sigma_N} + 2^{k-2} \sum_{y_1,\ldots,y_k} \left(\frac{1}{2} - \frac{F(y)}{2\sqrt{\pi}\sigma_N}\right)\right) \\
&= \frac{1}{4} \left(1 + \frac{\bar{g}(x_1, x_2) - \bar{g}}{\sqrt{\pi}\sigma_N}\right) ,
\end{aligned} \tag{4.34}$$

where $\bar{g}(x_1, x_2)$ denotes the average fitness contribution of the partition X with $X_1 = x_1$ and $X_2 = x_2$, and $\bar{g}$ denotes the average block fitness contribution of the partition X; that is,

$$\bar{g}(x_1, x_2) = \frac{1}{2^{k-2}} \sum_{x_3, \dots, x_k} g(x) \,, \tag{4.35}$$

and

$$\bar{g} = \frac{1}{2^k} \sum_{x_1, \dots, x_k} g(x) \,. \tag{4.36}$$

The dynamics of the two bits in a partition of order k can be accurately approximated by a special case of a two-bit partition with the fitness defined according to the average building-block fitnesses $\bar{g}(x_1, x_2)$. This property simplifies the derivations and allows the analysis of pairwise statistical dependencies independently of the order of the partition in which the two variables are located.

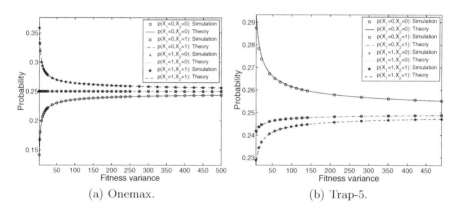

(a) Onemax.   (b) Trap-5.

Fig. 4.4. Pairwise frequencies versus their approximation with respect to collateral noise

The above approximations of the pairwise probabilities are verified in Fig. 4.4. The values computed according to the above approximations are compared to the actual values computed by the simulation of binary tournament selection using infinite populations on onemax and trap-5 functions. The simulation directly encodes all the assumptions and computes the probabilities without any approximations. In both cases, the results of the simulation and the approximations are practically indistinguishable.

4.3.6 General Two-Bit Case

In the general two-bit case, the fitness of the two variables X_1 and X_2 can be written as

$$\bar{g}(X_1, X_2) = a_0 + a_1 X_1 + a_2 X_2 + a_{12} X_1 X_2 , \tag{4.37}$$

where a_0, a_1, a_2, and a_{12} are constants. In the above equation, if the contribution of X_1 is independent of X_2, then $a_{12} = 0$. On the other hand, if $a_{12} \neq 0$, the contributions of the variables X_1 and X_2 are correlated. It is possible to create a problem that contains interactions of high order but where $a_{12} = 0$ for some pairs of variables within these interactions. For example, the fitness can be based on the parity of the bits involved in the partition of a building block. Nonetheless, this occurs only under special circumstances and in practice, this scenario is unlikely.

The probability of any instantiation of the two variables X_1 and X_2 after binary tournament selection can be computed using Equation (4.34), yielding

$$
\begin{aligned}
p(X_1, X_2) &= \frac{1}{4}\left(1 + \frac{\bar{g}(X_1, X_2) - \bar{g}}{\sqrt{\pi}\sigma_N}\right) \\
&= \frac{1}{4}\left(1 + \frac{a_0 + a_1 X_1 + a_2 X_2 + a_{12} X_1 X_2 - \frac{4a_0 + 2a_1 + 2a_2 + a_{12}}{4}}{\sqrt{\pi}\sigma_N}\right) \\
&= \frac{1}{4}\left(1 + \frac{a_1(4X_1 - 2) + a_2(4X_2 - 2) + a_{12}(4X_1 X_2 - 1)}{4\sqrt{\pi}\sigma_N}\right)
\end{aligned}
\tag{4.38}
$$

By summing the above equations over X_1 and X_2, respectively, we get

$$
\begin{aligned}
p(X_1) &= \frac{1}{2}\left(1 + \frac{a_1(4X_1 - 2) + a_{12}(2X_1 - 1)}{4\sqrt{\pi}\sigma_N}\right) \\
p(X_2) &= \frac{1}{2}\left(1 + \frac{a_2(4X_2 - 2) + a_{12}(2X_2 - 1)}{4\sqrt{\pi}\sigma_N}\right)
\end{aligned}
\tag{4.39}
$$

The above equations can be used to compute the frequencies of any instantiation of X_1 and X_2, yielding the following set of equations:

$$
\begin{aligned}
p(X_1 = 0, X_2 = 0) &= \frac{1}{4}\left(1 + \frac{-2a_1 - 2a_2 - a_{12}}{4\sqrt{\pi}\sigma_N}\right) = \frac{1}{4}(1 + \chi_{00}) \\
p(X_1 = 0, X_2 = 1) &= \frac{1}{4}\left(1 + \frac{-2a_1 + 2a_2 - a_{12}}{4\sqrt{\pi}\sigma_N}\right) = \frac{1}{4}(1 + \chi_{01}) \\
p(X_1 = 1, X_2 = 0) &= \frac{1}{4}\left(1 + \frac{2a_1 - 2a_2 - a_{12}}{4\sqrt{\pi}\sigma_N}\right) = \frac{1}{4}(1 + \chi_{10}) \\
p(X_1 = 1, X_2 = 1) &= \frac{1}{4}\left(1 + \frac{2a_1 + 2a_2 + 3a_{12}}{4\sqrt{\pi}\sigma_N}\right) = \frac{1}{4}(1 + \chi_{11})
\end{aligned}
\tag{4.40}
$$

In the last set of equations, the parameters χ_{ij} are defined to be equal to the terms they replace. Additionally, the probabilities of the instantiations of X_1 and X_2 can be computed as follows:

$$p(X_1 = 0) = \frac{1}{2}\left(1 + \frac{-2a_1 - a_{12}}{4\sqrt{\pi}\sigma_N^2}\right) = \frac{1}{2}(1 - \chi_1)$$

$$p(X_1 = 1) = \frac{1}{2}\left(1 + \frac{2a_1 + a_{12}}{4\sqrt{\pi}\sigma_N^2}\right) = \frac{1}{2}(1 + \chi_1)$$

$$p(X_2 = 0) = \frac{1}{2}\left(1 + \frac{-2a_2 - a_{12}}{4\sqrt{\pi}\sigma_N^2}\right) = \frac{1}{2}(1 - \chi_2)$$

$$p(X_2 = 1) = \frac{1}{2}\left(1 + \frac{2a_2 + a_{12}}{4\sqrt{\pi}\sigma_N^2}\right) = \frac{1}{2}(1 + \chi_2)$$

$$(4.41)$$

Again, the parameters χ_i are defined to be equal to the terms they replace.

Next, we compute the order of the growth of D in two separate cases. The first case considers two nonlinearly interacting variables where $a_{12} > 0$. The second case considers two variables whose contributions are independent and thus $a_{12} = 0$. In both the cases, the marginal entropies $H(X_1)$, $H(X_2)$, and $H(X_1, X_2)$ are first computed. These are then used to compute D, which can be expressed in terms of the marginal entropies as

$$D = H(X_1) - H(X_1|X_2) = H(X_1) + H(X_2) - H(X_1, X_2) . \qquad (4.42)$$

Dependent Case: $a_{12} > 0$

First, let us compute an approximation of the entropy of X_1 defined as

$$H(X_1) = -\sum_{x_1} p(x_1) \log_2 p(x_1) . \qquad (4.43)$$

Using the set of equations (4.41), the entropy of X_1 can be computed as

$$\begin{aligned}H(X_1) &= -\frac{1}{2}(1 - \chi_1)\log_2\frac{1}{2}(1 - \chi_1) - \frac{1}{2}(1 + \chi_1)\log_2\frac{1}{2}(1 + \chi_1) \\ &= -\frac{1}{2}\left(\log_2(1 - \chi_1^2) + \chi_1(\log_2(1 + \chi_1) - \log_2(1 - \chi_1))\right) + 1\end{aligned} \qquad (4.44)$$

Since χ_1 is very small for moderate-to-large sized problems (it approaches zero as σ_N approaches infinity), we can use a linear approximation of the logarithm near 1. In particular,

$$\log_2(1 - \chi_1^2) \approx -\frac{\chi_1^2}{\ln 2} ,$$

$$\chi_1 \log_2(1 + \chi_1) \approx \frac{\chi_1^2}{\ln 2} , \qquad (4.45)$$

$$\chi_1 \log_2(1 - \chi_1) \approx -\frac{\chi_1^2}{\ln 2} .$$

Thus,

$$H(X_1) = -\frac{\chi_1^2}{2\ln 2} + 1$$
$$= -\frac{4a_1^2 + 4a_1 a_{12} + a_{12}^2}{32\pi\sigma_N^2 \ln 2} + 1 \ . \tag{4.46}$$

$H(X_2)$ can be computed analogously to $H(X_1)$, yielding

$$H(X_2) = -\frac{4a_2^2 + 4a_2 a_{12} + a_{12}^2}{32\pi\sigma_N^2 \ln 2} + 1 \ .$$

The joint entropy $H(X_1, X_2)$ is given by

$$H(X_1, X_2) = -\sum_{x_1, x_2} p(x_1, x_2) \log_2 p(x_1, x_2) = -(A_{00} + A_{01} + A_{10} + A_{11}) \ ,$$

where

$$A_{ij} = p(X_1 = i, X_2 = j) \log_2 p(X_1 = i, X_2 = j) \ . \tag{4.47}$$

The terms A_{ij} can be approximated as follows:

$$A_{ij} = \frac{1}{4}\left((1 + \chi_{ij}) \log_2 \frac{1}{4}(1 + \chi_{ij}) \right)$$
$$= \frac{1}{4}\left(\log_2(1 + \chi_{ij}) + \chi_{ij} \log_2(1 + \chi_{ij}) - 2(1 + \chi_{ij}) \right) \tag{4.48}$$

Since χ_{ij} is very small, the last equation can be simplified using the following approximations:

$$\log_2(1 + \chi_{ij}) \approx \frac{2\chi_{ij} - \chi_{ij}^2}{2\ln 2} \ ,$$
$$\chi_{ij} \log_2(1 + \chi_{ij}) \approx \frac{\chi_{ij}^2}{\ln 2} \ . \tag{4.49}$$

Thus,

$$A_{ij} = \frac{1}{4}\left(\frac{2\chi_{ij} + \chi_{ij}^2}{2\ln 2} - 2(1 + \chi_{ij}) \right) \ . \tag{4.50}$$

By substituting the approximations of A_{ij} and the equations for χ_{ij}, we get

$$H(X_1, X_2) = -\frac{1}{8}\left(\frac{16a_1^2 + 16a_2^2 + 12a_{12}^2 + 16a_1 a_{12} + 16a_2 a_{12}}{16\pi\sigma_N^2 \ln 2} \right) + 2 \tag{4.51}$$

Thus, the difference D between the marginal and conditional entropies can be approximated by

$$D = \frac{a_{12}^2}{32\pi\sigma_N^2 \ln 2} \ . \tag{4.52}$$

Therefore, if the fitness contributions of X_1 and X_2 are not independent, then D grows inversely proportionally to the variance of collateral noise.

By substituting Equation (4.52) into Equation (4.25), we can infer that

$$N_{crit} = O(\sigma_N^{2.1}) . \tag{4.53}$$

Using the assumption that $\sigma_N^2 \propto n$ where n is the number of variables in the problem, we can imply that

$$N_{crit} = O(n^{1.05}) . \tag{4.54}$$

In other words, the critical population size for discovering a dependency between the two variables that are nonlinearly correlated grows approximately linearly with problem size.

Independent Case: $a_{12} = 0$

Note that in this case, $a_{12} = 0$. Therefore,

$$\chi_1 = \frac{a_1}{2\sqrt{\pi}\sigma_N} , \quad \chi_2 = \frac{a_2}{2\sqrt{\pi}\sigma_N} . \tag{4.55}$$

We can now write χ_2 in terms of χ_1 as

$$\chi_2 = \frac{a_2}{a_1}\chi_1 = b\chi_1 , \tag{4.56}$$

where $b = a_2/a_1$. The entropy of X_i (here we're interested in X_1 and X_2 only) can be written as

$$
\begin{aligned}
H(X_i) &= -\frac{1}{2}\left((1 + \chi_i)\log_2\left(\frac{1 + \chi_i}{2}\right) + (1 - \chi_i)\log_2\left(\frac{1 - \chi_i}{2}\right) \right) \\
&= -\frac{1}{2}\left((1 + \chi_i)\log_2(1 + \chi_i) + (1 - \chi_i)\log_2(1 - \chi_i) - 2 \right) \\
&= -\frac{1}{2}\left(\log_2((1 + \chi_i)(1 - \chi_i)) + \chi_i\log_2\left(\frac{1 + \chi_i}{1 - \chi_i}\right) - 2 \right) . \tag{4.57}
\end{aligned}
$$

Since χ_i is very small, the logarithms in the last equation can be approximated as

$$\log_2((1 + \chi_i)(1 - \chi_i)) \approx -\frac{2\chi_i^2 + \chi_i^4}{2\ln 2} , \tag{4.58}$$

$$\log_2\left(\frac{1 + \chi_i}{1 - \chi_i}\right) \approx \frac{6\chi_i + 2\chi_i^3}{3\ln 2} . \tag{4.59}$$

Using the above approximations, the entropy $H(X_i)$ is given by

$$H(X_i) = -\frac{1}{2}\left(\frac{6\chi_i^2 + \chi_i^4}{6\ln 2} - 2 \right) . \tag{4.60}$$

Thus, the entropies $H(X_1)$ and $H(X_2)$ are given by

$$H(X_1) = -\frac{6\chi_1^2 + \chi_1^4}{12\ln 2} + 1 \ . \tag{4.61}$$

$$H(X_2) = -\frac{6\chi_2^2 + \chi_2^4}{12\ln 2} + 1 \ ,$$

$$= -\frac{6b^2\chi_1^2 + b^4\chi_1^4}{12\ln 2} + 1 \ . \tag{4.62}$$

The only other term remaining to compute the entropy difference D is the joint entropy, $H(X_1, X_2)$, which is given by

$$H(X_1, X_2) = -\left(A_{00} + A_{01} + A_{10} + A_{11}\right) \ ,$$

where

$$A_{ij} = \frac{1}{4}\left((1 + \chi_{ij})\log_2\left(\frac{1 + \chi_{ij}}{4}\right)\right)$$

$$= \frac{1}{4}\left(\log_2\left(1 + \chi_{ij}\right) + \chi_{ij}\log_2\left(1 + \chi_{ij}\right) - 2\left(1 + \chi_{ij}\right)\right) \ , \tag{4.63}$$

and

$$\begin{aligned}
\chi_{00} &= -(1 + b)\chi_1 \ , \\
\chi_{01} &= -(1 - b)\chi_1 \ , \\
\chi_{10} &= (1 - b)\chi_1 \ , \\
\chi_{11} &= (1 + b)\chi_1 \ .
\end{aligned} \tag{4.64}$$

From the above equation, we get

$$\begin{aligned}
A_{00} &= \frac{1}{4}\left(\log_2\left(1 - (1 + b)\chi_1\right) - (1 + b)\chi_1\log_2\left(1 - (1 + b)\chi_1\right)\right. \\
&\quad \left. -2\left(1 - (1 + b)\chi_1\right)\right) \ , \\
A_{01} &= \frac{1}{4}\left(\log_2\left(1 - (1 - b)\chi_1\right) - (1 - b)\chi_1\log_2\left(1 - (1 - b)\chi_1\right)\right. \\
&\quad \left. -2\left(1 - (1 - b)\chi_1\right)\right) \ , \\
A_{10} &= \frac{1}{4}\left(\log_2\left(1 + (1 - b)\chi_1\right) + (1 - b)\chi_1\log_2\left(1 + (1 - b)\chi_1\right)\right. \\
&\quad \left. -2\left(1 + (1 - b)\chi_1\right)\right) \ , \\
A_{11} &= \frac{1}{4}\left(\log_2\left(1 + (1 + b)\chi_1\right) + (1 + b)\chi_1\log_2\left(1 + (1 + b)\chi_1\right)\right. \\
&\quad \left. -2\left(1 + (1 + b)\chi_1\right)\right) \ .
\end{aligned}$$

Summing the above equations for A_{ij} gives us $H(X_1, X_2)$:

$$\begin{aligned}
H(X_1, X_2) = -\frac{1}{4}\left(\log_2\left((1 - (1 + b)^2\chi_1^2)(1 - (1 - b)^2\chi_1^2)\right)\right. \\
+ (1 + b)\chi_1\log_2\left(\frac{1 + (1 + b)\chi_1}{1 - (1 + b)\chi_1}\right) \\
\left. + (1 - b)\chi_1\log_2\left(\frac{1 + (1 - b)\chi_1}{1 - (1 - b)\chi_1}\right) - 8\right) \ . \tag{4.65}
\end{aligned}$$

Using the approximations from equations (4.58) and (4.59),

$$
\begin{aligned}
H(X_1, X_2) &= -\frac{1}{4\ln 2}\left((1+b)^2\chi_1^2 + \frac{1}{6}(1+b)^4\chi_1^4 + (1-b)^2\chi_1^2\right.\\
&\quad\left. +\frac{1}{6}(1-b)^4\chi_1^4\right) + 2\\
&= -\frac{1}{4\ln 2}\left(\left((1+b)^2 + (1-b)^2\right)\chi_1^2 + \frac{1}{6}((1+b)^4\right.\\
&\quad\left. + (1-b)^4)\chi_1^4\right) + 2\\
&= -\frac{1}{2\ln 2}\left(1+b^2\right)\chi^2 \qquad\qquad (4.66)\\
&\quad -\frac{1}{12\ln 2}\left(1+6b^2+b^4\right)\chi_1^4 + 2 .
\end{aligned}
$$

The entropy difference D can be computed by substituting equations (4.61), (4.62), and (4.66) into Equation (4.42):

$$
\begin{aligned}
D &= \frac{1}{2\ln 2}\left(1+b^2\right)\chi^2 + \frac{1}{12\ln 2}\left(1+6b^2+b^4\right)\chi^4 - \frac{1}{2\ln 2}\left(1+b^2\right)\chi^2\\
&\quad -\frac{1}{12\ln 2}\left(1+b^4\right)\chi^4\\
&= \frac{b^2}{2\ln 2}\chi_1^4 . \qquad\qquad (4.67)
\end{aligned}
$$

Recall that $\chi_1 = \frac{a_1}{2\sqrt{\pi}\sigma_N}$, and $b = a_2/a_1$. Therefore, the above equation can be written in terms of the variance of collateral noise σ_N^2 as

$$
D = \frac{1}{32\ln 2}\left(\frac{a_1 a_2}{\pi}\right)^2 \sigma_N^{-4} . \qquad\qquad (4.68)
$$

By substituting Equation (4.68) into Equation (4.25), we can imply that if $a_{12} = 0$, then

$$
N_{crit} = O(\sigma_N^{4.2}) , \qquad\qquad (4.69)
$$

which yields

$$
N_{crit} = O(n^{2.1}) . \qquad\qquad (4.70)
$$

In other words, the population size to discover the dependency between the two variables that are independent with respect to the fitness function grows approximately quadratically with the problem size.

The above theory assumes that X_1 has no parents before the decision is made on whether the edge $X_2 \rightarrow X_1$ should be added into the network (see Fig. 4.2(a)). The following section discusses an extension of the theory to the general case where X_1 already has a number of parents before the decision regarding the edge $X_2 \rightarrow X_1$ is made (see Fig. 4.2(b)). Subsequently, the section justifies the assumption that the frequencies follow their expected behavior by incorporating the effects of a finite population sizing into the model.

4.3.7 General Case: Multiple Parents of X_1 Exist

Above we computed the required population size for the addition of the first edge into X_1. How does the situation change if some edges that end in X_1 are already present in the current model? This section indicates that even in that case, the overall growth of the population size for a constant order of interactions does not change much, although the population size must grow exponentially with the order of the considered dependencies.

The condition for adding the edge $X_2 \rightarrow X_1$ into the network that already contains the edges $X_3 \rightarrow X_1$ to $X_k \rightarrow X_1$ is given by

$$BIC(X_1|X_2,\ldots,X_k) > BIC(X_1|X_3,\ldots X_k) . \qquad (4.71)$$

Using the definition of BIC, the last equation can be rewritten as

$$- H(X_1|X_2,\ldots,X_k)N - 2^{k-2}\log_2 N > -H(X_1|X_3,\ldots,X_k) - 2^{k-3}\log_2 N . \qquad (4.72)$$

Denoting $D = H(X_1|X_3\ldots X_k) - H(X_1|X_2\ldots X_k)$ yields

$$N - \frac{\log_2 N}{2^{3-k}D} > 0 . \qquad (4.73)$$

Analogously to the case of the first-edge addition, the critical population size is the larger of the two solutions of the above equation and the growth of D determines the growth of the critical population size.

To determine the growth of D, let us first discuss the form of the nonlinearities that should be discovered in this case. If the contribution of X_1 does not depend on X_2 given the values of X_3 to X_k, then the edge $X_2 \rightarrow X_1$ is not required, because there is no additional nonlinearity that must be covered in the model. However, if there is some combination of values of X_3 to X_k for which the contributions of X_1 and X_2 are correlated, the edge $X_2 \rightarrow X_1$ should be added to reflect the nonlinearity.

The discussion in the above paragraph suggests that the nonlinearities that are *conditioned* on the particular values of X_3 to X_k are important to cover. In that case, the growth of D can be approximated by partitioning the population according to the instantiations of X_3 to X_k, and looking at each subpopulation separately. Note that the size of each partition of the population is approximately $N/2^{k-2}$, because the probabilities of the blocks of $k - 2$ bits get closer to each other asymptotically. Figure 4.5 shows an example partitioning of the population according to the first 3 bits.

Using the subpopulations of the partitioning according to $(X_3,\ldots,X_k)$, the overall D can be computed as the weighted sum of D's for each subpopulation, because

$$H(X_1|X_3,\ldots,X_k) = \sum_{x_3,\ldots,x_k} p(x_3,\ldots,x_k)H_{x_3,\ldots,x_k}(X_1) , \qquad (4.74)$$

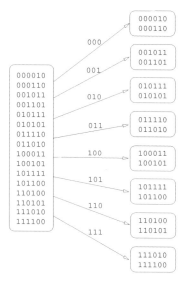

Fig. 4.5. Partitioning the population according to the first three bits. For each configuration of the first three bits, a subpopulation is created that consists of all the individuals that contain that configuration

and

$$H(X_1|X_2, X_3, \ldots, X_k) = \sum_{x_3,\ldots,x_k} p(x_3,\ldots,x_k) H_{x_3,\ldots,x_k}(X_1|X_2) , \qquad (4.75)$$

where $H_{x_3,\ldots,x_k}(X)$ denotes the entropy of X in the partition of $(x_3, \ldots, x_k)$. Consequently, if the fitness contributions of X_1 and X_2 are correlated in at least one partition, the dominant term in D will be the one coming from that partition, and the dependency will be found if the size of that partition grows as $O(n^{1.05})$. Since the expected size of each partition is $N/2^{k-2}$, the overall growth of the population size can be bounded by $O(2^{k-2}n^{1.05})$. If the contributions of X_1 and X_2 are correlated in more than one partition, the population-sizing bound can be decreased accordingly. On the other hand, if the contributions of X_1 and X_2 are independent in every context of X_3 to X_k, the population size to discover this unnecessary dependency $X_2 \rightarrow X_1$ should grow as $O(2^{k-2}n^{2.1})$.

Therefore, the theory for the case of adding the first edge that ends in X_1 can be extended to the general case in a straightforward manner, yielding the overall bound on the population size of

$$N_{crit} = O(2^k n^{1.05}) , \qquad (4.76)$$

where k is the maximum order of the subproblems in the problem decomposition. The above result has two important implications:

(1) **Sufficient population size.** The sufficient population size for discovering the nonlinearities in the problem and encoding them in the learned Bayesian network grows approximately with $O(2^k n^{1.05})$. Assuming the fixed order of decomposition, the growth is $O(n^{1.05})$.

(2) **Favored nonlinearities.** The magnitude of the nonlinearities affects the population sizing. The higher the magnitude of the nonlinearities, the smaller the population size. Assuming a particular population size, only the strongest nonlinearities can be covered, because the population size grows exponentially with the order of the covered dependencies.

Both the dependent case and the independent one assume that the frequencies after selection are equal to their expected values. The following section analyzes the effects of using finite populations on the accuracy of the actual frequencies and incorporates the results of the analysis into the developed model.

4.3.8 Getting the Frequencies Right

The population-sizing model presented above assumes that the frequencies after binary tournament selection are equal to their expected values. However, in practice we can only use finite populations and the model should also consider the effects of the finite population sizing on the actual, observed, frequencies and, consequently, on the model building in BOA. The purpose of this section is to analyze the effects of the finite population sizing and apply the results of the analysis to the presented population-sizing model. In particular, a lower bound on the population size is computed that ensures that the actual frequencies of each block of k bits are close enough to their expected values with high confidence, where "close enough" will be defined later.

Assume that the probability of x being a winner of one tournament is equal to its expected probability $p(x)$ after selection. Let us denote the actual probability (relative frequency) of x in the selected set of solutions after performing m tournaments by $p_m(X)$. Note that after m tournaments there are m solutions selected (m is the size of the selected population) and, since the tournaments are stochastic, $p_m(X)$ is a random variable.

The distribution of $p_m(X)$ is binomial, because $p_m(X)$ is equal to the number of successes in m independent trials divided by the number of trials, each trial with the probability of success equal to $p(x)$. The mean of $p_m(x)$ is $p(x)$ and the variance of $p_m(x)$ is $p(x)(1 - p(x))/m$:

$$p_m(x) \sim Bin\left(p(x), \frac{p(x)\,(1 - p(x))}{m}\right) . \tag{4.77}$$

For moderate values of m the binomial distribution can be approximated by the normal distribution, yielding

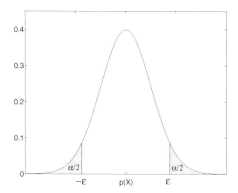

Fig. 4.6. To ensure that $p_m(X)$ is within ϵ of its mean, the areas under the tails that start at distance ϵ from the mean must sum to α

$$p_m(X) \sim N\left(p(X), \frac{p(X)\,(1-p(X))}{m}\right) . \tag{4.78}$$

With confidence α the actual frequency $p_m(X)$ is within ϵ from its expected value $p(X)$, if

$$\Phi\left(-\frac{\epsilon\sqrt{m}}{\sqrt{p(X)(1-p(X))}}\right) \leq \frac{\alpha}{2} , \tag{4.79}$$

where Φ is the cumulative density of the unit normal distribution (see Fig. 4.6). For m, we get

$$m \geq \left(\frac{\Phi^{-1}\left(\frac{\alpha}{2}\right)\sqrt{p(X)(1-p(X))}}{\epsilon}\right)^2 , \tag{4.80}$$

where Φ^{-1} is the inverse cumulative density of the unit normal distribution. For constant confidence α, the number of tournaments must therefore grow inversely proportionally with the square of ϵ.

How should we set the value of ϵ for the population sizing in BOA? Let us first get back to the block probabilities and their dynamics with the problem size. The frequencies of any two-bit partition approach 0.25 inversely proportionally to the standard deviation σ_N of collateral noise. It would be therefore reasonable to set the error ϵ to decrease at the same rate so that the same relative accuracy could be achieved for the entire spectrum of problem sizes. For instance, the distance of the frequencies to their asymptotic value could deviate by at most 1% independently of the size of the problem. In that case, the actual population size could be bounded by the two extreme cases at an arbitrary level of confidence.

The frequencies for bigger partitions behave similarly, but they are scaled down by an additional factor of 2^{k-2} where k is the order of the considered partition (see Equation (4.32)). So the accuracy of the frequencies should also increase proportionally to 2^k. Therefore, it is reasonable to require that

$$\epsilon \propto \frac{1}{2^k \sigma_N} \, , \tag{4.81}$$

where k is the size of the building blocks we must consider to find the optimum. Using the assumption that $\sigma_N^2 \propto n$ (see Equation (4.18)), we get

$$\epsilon \propto \frac{1}{2^k \sqrt{n}} \, , \tag{4.82}$$

where n is the size of the problem. Substituting the last equation into Equation (4.80) yields

$$m = O(2^k n) \, . \tag{4.83}$$

For large tournament sizes, the probability of a solution x being the winner of a random tournament might differ from its expected value $p(x)$. Nonetheless, by enlarging the population by a factor equal to the tournament size, this assumption can be justified even for large tournament sizes.

Therefore, for a constant bound k on the order of the subproblems, the population size to ensure that the frequencies retain the same relative error with arbitrary confidence grows linearly with the problem size. Since the growth of the population sizes in both the dependent case and the independent one was at least linear as well, the population-sizing model for the BIC metric is applicable to the case of finite populations.

The following section presents empirical results that verify the model and the approximations made in its derivation.

4.3.9 Critical Population Size: Empirical Results

Figure 4.7(a) shows the critical population size for onemax (see Equation (1.1) on page 5). Both the simulation for infinite populations (based on the exact theoretical model for the frequencies using an infinite population) as well as the final approximate result are shown. We can see that the match between the theory and the infinite-population simulation is very good and that the critical population size for discovering a dependency between the variables whose fitness contributions are independent increases approximately quadratically with the fitness variance, which is proportional to the size of the problem.

Figure 4.7(b) shows the critical population size for trap-5 compared to the empirical results for the simulation with finite and infinite populations. The correlated bits are both selected from one of the trap subfunctions, whereas the independent bits are selected from two different subfunctions. The infinite-population simulation was again performed according to the assumptions stated in Sect. 4.3.3. The simulation for a finite population was performed by simulating the actual binary tournament selection on a finite population and increasing the population size until the considered dependency was discovered in at least 95 out of 100 independent runs. The exact theoretical results and the approximations match very well. We can also see that the use of a

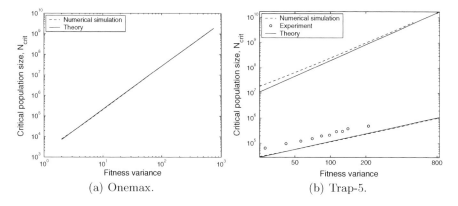

Fig. 4.7. Critical population size for onemax and trap-5 using BIC metric

finite population introduces additional noise that increases the population-sizing requirements for a reliable detection of the correct dependencies, but that the growth of the appropriate population size is still approximately linear.

The results in Fig. 4.7(b) also indicate that there is a large range of population sizes that result in a reliable discovery of nonlinear dependencies but still do not introduce unnecessary dependencies between the variables whose fitness contributions are independent. Moreover, the range grows with the problem size.

Our theoretical analysis considered only *binary* tournament selection. Figure 4.8(a) indicates that the range of population sizes leading to the discovery of good dependencies but ensuring that the algorithm is not misled by the bad dependencies grows with the selection pressure. The bad news is that the growth of the required population sizes grows slightly faster with an increased selection pressure. For the tournament size of $s = 2$, the actual growth of the population size is approximately $O(n^{1.035})$. For the tournament size of $s = 16$, the growth increases to $O(n^{1.242})$. On the other hand, the order of the growth of the population size required to discover the bad dependencies decreases from 1.974 for $s = 2$ to 1.572 for $s = 16$. Nonetheless, as mentioned above, the range of adequate population sizes still increases with the selection pressure.

Figure 4.8(b) shows that increasing the tournament size up to $s = 16$ decreases the critical population size even for the case of a finite population. However, for high selection pressures, the positive effects of increasing the selection pressure can be expected to decrease and actually harm the performance of the algorithm in practice due to the premature convergence.

We observed similar results regarding the discovery of dependencies of higher order for both the onemax and trap-5 functions.

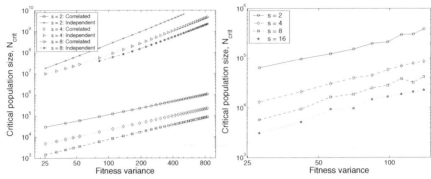

(a) Simulation with infinite population.   (b) Experiment with finite population.

Fig. 4.8. The effects of increasing the selection pressure on the critical population size. As the selection pressure increases, the critical population size decreases. The reason for this behavior is that increasing the selection pressure results in increasing the effect of each nonlinearity on the frequencies after selection

4.3.10 Summary of Population Sizing in BOA

This section summarizes the results of the aforementioned population-sizing models for GAs and BOA.

The requirements on the population size in BOA due to the four factors of the BOA population sizing (see Sect. 4.3) are listed in Table 4.2. For uniformly scaled subproblems and a bounded size of the subproblems, the population-sizing requirements for the BIC metric to distinguish between the good and the bad dependencies, are dominant. The population size must grow as

$$N_{uni} = O(n^{1.05}) . \tag{4.84}$$

For exponentially scaled problems, the model building is easy because the entire population is used to determine the model over only a small subset of

Table 4.2. An overview of the factors influencing the population sizing in BOA and the corresponding bounds. The first three factors are adopted from the GA population sizing theory. The last factor is introduced to ensure that BOA builds an adequate model

Factor	Population Size
Initial supply	$O(2^k(k + \log n))$
Decision making	$O(2^k \sqrt{n})$
Genetic drift	$O(\sqrt{n})$ to $O(n)$
Model building	$O(2^k n^{1.05})$ for putting in the right edges, or $O(2^k n)$ for getting the frequencies right.

the variables that are being optimized at a time. The effects of genetic drift become the dominant factor, yielding

$$N_{exp} = O(n) \ . \tag{4.85}$$

The following section provides background of GA theory for estimating the number of generations until convergence. Section 4.5 relates the GA time-to-convergence theory to BOA. Subsequently, Sect. 4.6 combines the BOA population-sizing theory with the estimated number of generations until convergence to compute the total number of evaluations.

4.4 Background of GA Time-to-Convergence Theory

Once we have an adequate population size to find a solution of a specified quality, another important question is how many generations it will take the algorithm to converge. As discussed above, it is important to distinguish between the problems with uniform and exponential scaling. The number of generations until convergence in the exponential case was discussed above in the context of the drift population-sizing model and the interested reader should refer to the work cited in the previous section. This section focuses on the uniformly scaled subproblems.

Mühlenbein and Schlierkamp-Voosen [118] estimated the number of generations until convergence on onemax, assuming an infinite population size and perfect mixing using population-wise uniform crossover, as follows:

$$t_{conv} = \left(\frac{\pi}{2} - \arcsin\left(2p - 1\right) \right) \frac{\sqrt{n}}{I} \ , \tag{4.86}$$

where p is the initial proportion of ones on each position; n is the problem size; and I is the selection intensity. Selection intensity [16] is defined as the difference between the mean fitness of the population before selection and the population after selection normalized by the standard deviation of the fitness values in the population before selection:

$$I(t) = \frac{\overline{f}(t+1) - \overline{f}(t)}{\sigma(t)} \ , \tag{4.87}$$

where $\overline{f}(t+1)$ is the average fitness of the population in generation $t+1$; $\overline{f}(t)$ is the average fitness in generation t; and $\sigma(t)$ is the standard deviation of the fitness values in generation t. For many commonly used selection schemes – such as the tournament, truncation, ranking, and (λ, μ) selection – the selection intensity is a constant [18, 118, 173].

The above result suggests that with a large enough population, the number of generations until convergence is proportional to the square root of the problem size and inversely proportional to the selection intensity. The above convergence model was later rederived for tournament selection and extended by Miller and Goldberg [107] to take into account additional normally distributed noise in the fitness function. Finally, Bäck [5] developed the convergence model for (λ, μ) selection.

The above two convergence models assume that the population is large enough for the convergence to follow its expected behavior with an infinite population size. In practice, the population size must usually be sufficiently large for the convergence to the optimum; in that case the GA dynamics is indeed close to the case with infinite populations. However, the dynamics changes if the populations are much smaller, which is often the case in parallel GA implementations that distribute the population on several processors and evolve each subpopulation in isolation. Ceroni et al. [20] incorporated the effects of small populations on the overall number of generations until convergence. The primary use of the convergence model of Ceroni et al. [20] is in maximizing the speed-up of parallel GAs.

The following section considers the number of generations until convergence in BOA.

4.5 Time to Convergence in BOA

The last piece we must collect to complete the puzzle of BOA scalability is the time to convergence. The following two sections compute the number of generations until BOA convergence in two bounding cases: (1) uniformly scaled subproblems, and (2) exponentially scaled subproblems. In the first case, the fitness contributions of all the subproblems of the decomposition are scaled the same and all the partitions converge in parallel [5, 107, 118, 175]. In the second case, the subproblems can be ordered so that the fitness contributions of any subproblem overshadow the contributions of the remaining subproblems in the ordering; in this case, the partitions corresponding to the different subproblems converge sequentially, one partition after another, in a domino-like fashion [177].

4.5.1 Uniform Scaling

The number of generations until convergence in BOA can be modeled analogously to onemax and population-wise uniform crossover. In that case, Mühlenbein and Schlierkamp-Voosen [118] showed that the number of generations until convergence grows as

$$G_{onemax} = \left(\frac{\pi}{2} - \arcsin\left(2p - 1\right) \right) \frac{\sqrt{n}}{I} , \tag{4.88}$$

where p is the proportion of ones on each position in the initial generation, n is the problem size, and I is the selection intensity (see Equation (4.87)). For most commonly used selection methods – such as tournament and truncation selection – the selection intensity is constant and the number of generations is therefore bounded by

$$G_{onemax} = O(\sqrt{n}) \ . \tag{4.89}$$

Although the exact approximation of G given in Equation (4.88) is correct only for a simple model with no interactions applied to the onemax case, the model can be used to accurately model the convergence time of BOA on many other decomposable problems where the order of each subproblem is bounded by a constant and the contributions of all the subproblems are scaled the same [107, 131]. When the dynamics of the fitness variance is similar to the onemax case, Equation (4.88) approximates the time to convergence very well. Even if this is not the case, the time to convergence can still be accurately approximated by fitting G according to Equation (4.89). The reason for that behavior is that the time convergence can be upper-bounded by the the number of generations it would take to converge if the initial population contained only two partial solutions in each partition – the building block and its toughest competitor. In this case, the assumptions of the onemax convergence model hold if an adequate probabilistic model is used, and the number of generations can be computed using Equation (4.88).

Figure 4.9 (from Pelikan et al. [131]) shows the number of generations until convergence of BOA with truncation selection with $s = 2$, which selects the best half of the population in each generation. The empirical results on trap-5 and onemax fitness functions are compared to the prediction according to Equation (4.88). In both cases, the results match the original approximation for onemax with the model that contains no interactions.

The above theory assumes that the population is sufficiently large. To incorporate the effects of rather small populations, the convergence model of Ceroni et al. [20] can be adopted in a similar fashion. Additionally, the above

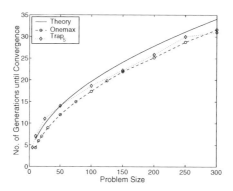

Fig. 4.9. The number of generations until convergence of BOA on onemax and trap-5. The problem sizes range from 0 to 300 bits. Large populations are used to approximate the asymptotic behavior

model assumes no external noise in the fitness function. To incorporate the effects of external noise, the noisy convergence model of Miller and Goldberg [107] can be used.

4.5.2 Exponential Scaling

The above convergence model assumed that all the subproblems converge in parallel. Under that assumption, the number of generations until convergence is $O(\sqrt{n})$. However, if the scaling of the subproblems is such that the subproblems converge sequentially, the time to convergence can further increase [2, 101, 152, 177].

Let us get back to the example problem of exponentially scaled subproblems – the binary integer (see Equation (4.7)). The binary integer converges one or a few bits at a time, and its overall time to convergence is therefore proportional to the number of bits [177]:

$$G_{binInt} = O(n) , \qquad (4.90)$$

where n is the number of bits in the problem (problem size).

Since number of generations until convergence increases with the number of phases (different levels of scaling), the linear time gives us an upper bound on the time to convergence for decomposable problems of bounded difficulty. Regardless of the scaling of the different building blocks in the problem, the time to convergence should be somewhere between $O(\sqrt{n})$ and $O(n)$.

4.6 How does BOA Scale Up?

Section 4.1 has argued that the number of fitness evaluations is an important quantity for determining the overall computational complexity of BOA. The section has stated that the number of evaluations can be bounded by a product of the population size and the number of generations until convergence. Boundary cases have been then analyzed and the bounds on both the population size as well as the number of generations until convergence have been presented. This section puts all the pieces of the theory together and computes the overall number of function evaluations until convergence to the optimum.

The overall result is summarized in Table 4.4. Under the assumption that the problem is decomposable into the subproblems of bounded order, there are two bounding cases. One one hand, if the subproblems in the decomposition are scaled the same, the total number of evaluations is given by

$$E_{uni} = O(n^{1.55}) . \qquad (4.91)$$

On the other hand, if the subproblems are scaled exponentially, the total number of fitness evaluations is given by

$$E_{exp} = O(n^2) . \tag{4.92}$$

In both cases, the total number of evaluations is bounded by a low-order polynomial of the problem size.

Let us now compare the above result to the local search; more specifically, let us consider a stochastic hill climber with bit-flip mutation described in Sect. 3.5.2. The hill-climber requires $O(n^k \ln n)$ fitness evaluations for the problem decomposable into subproblems of order k [112]. While in BOA the difficulty of the problem (expressed in the form of k) is hidden in the constants preceding a polynomial of the fixed order, in the hill climber the difficulty directly affects the order of the growth of the number of evaluations. Although for $k = 1$ or $k = 2$ this does not make a big difference, for moderate values of k the difference becomes the one between the tractable and the intractable (recall the comparison presented in the previous chapter in Figs. 3.12 and 3.13).

To get a better idea of the difference between BOA and the hill climber, consider the following example (see Table 4.3 for a summary of the example). Assume that both algorithms are capable of solving the problem of $n = 100$ bits decomposable into subproblems of order $k = 5$ (e.g., trap-5). To solve the same problem of twice the size, $n = 200$, the number of evaluations required by BOA would increase about 2.93 times, whereas the number of evaluations required by the hill climber would increase 32 times. If we were to solve the problem of $n = 300$ bits, the number of evaluations compared to the problem of $n = 100$ bits for BOA would be larger by a factor of 5.49, whereas for the hill climber it would be 243. The difference further increases with the problem size. For instance, for the problem of $n = 500$ bits, the factor is only 12.12 for BOA but it is $3,125$ for the hill climber.

Comparison of BOA and GAs with common crossover operators, such as one-point and uniform crossover, leads to even more dramatic differences as was indicated by the results presented in the previous chapter. It is known that GAs with common crossover operators may require exponentially many evaluations for solving decomposable problems of bounded difficulty [57, 172]. That

Table 4.3. The factor by which the number of evaluations increases in BOA and the hill climber (HC) on a problem decomposable into uniformly scaled subproblems of order $k = 5$ (e.g., trap-5). Compared to the problem of $n = 100$ bits, the number of evaluations required by BOA to solve the problem of $n = 500$ bits increases approximately 12.12 times, whereas the number of evaluations required by the stochastic hill climber increases $3,125$ times

Problem Size	BOA	HC
$n = 100$	1.00	1.00
$n = 200$	2.93	32.00
$n = 300$	5.49	243.00
$n = 400$	8.57	1,024.00
$n = 500$	12.12	3,125.00

Table 4.4. A summary of the total number of evaluations in BOA. Assuming a fixed order of problem decomposition, the overall number of fitness evaluations is expected to grow subquadratically or quadratically with the size of the problem

	Uniform Scaling	Exponential Scaling
Population size	$O(n^{1.05})$	$O(n)$
Time to convergence	$O(\sqrt{n})$	$O(n)$
Total evaluations	$O(n^{1.55})$	$O(n^2)$

is why the search with standard GAs becomes intractable for moderate-to-large problems except for problems where an adequate problem decomposition directly corresponds to the used crossover operator.

The situation becomes somewhat more complicated if the order of the building blocks in the problem is not bounded by a constant but grows with the problem size instead. It can be argued that if the character of the sub-problems does not change much and their order grows logarithmically with the problem size, the performance further increases to a polynomial of higher order, because of the terms 2^k in the population-sizing bounds. Further refinement and generalization of the given theory remain for future research.

The following section verifies the scalability estimates presented above on several problems of bounded difficulty.

4.7 Empirical Verification of BOA Scalability

This section verifies the above theoretical bounds on the population size and the number of generations until convergence with empirical results. Experiments are divided into two parts. The first part considers uniformly scaled problems, whereas the second one considers exponentially scaled problems.

Experiments are designed similarly as in the previous chapter. For each problem and each problem size, the population size is determined by the bisection method as the minimum population size to ensure that BOA finds the optimum in 30 independent runs (see Fig. 3.8 for the pseudocode of the bisection method). Binary tournament selection without replacement is used. The worst half of the original population is replaced by the same number of offspring. BIC metric is used to evaluate candidate models. In each generation, the model is created from scratch. All results are averaged over the 30 independent runs with the minimum population size.

4.7.1 Uniform Scaling

To verify the scalability results for uniform problems, BOA was tested on onemax, trap-5, and deceptive-3 functions. Onemax is defined as the sum of bits in the input string (see Sect. 1.4). Trap-5 is defined as the sum of single trap functions of order 5 applied to non-overlapping 5-bit partitions of solution strings (see Sect. 1.5 for a detailed definition). The partitioning is fixed but BOA is given no information about the positions in each partition. Deceptive-3 [129] is defined as the sum of deceptive functions of order 3 applied to non-overlapping 3-bit partitions of solution strings (see Sect. 3.5.1 for a detailed definition). The partitioning is fixed but BOA is given no information about the positions in each partition.

On onemax, the model learned by BOA does not have to encode any interactions. Therefore, the model-building population sizing does not have to be considered and the expected performance of BOA should be close to that of the simple GA with population-wise uniform crossover. The population sizing should follow the gambler's ruin population-sizing model [70], and the number of generations should follow the onemax convergence model for population-wise uniform crossover [118]. Therefore, the total number of evaluations should grow as $O(n \log n)$. Figure 4.10 shows the total number of evaluations until the convergence of BOA on onemax. The results match the theory well. However, due to the errors introduced by using finite populations, BOA finds some unnecessary dependencies; consequently, BOA processes larger partitions than necessary, and its performance increases compared to the simple GA with uniform crossover, which represents an ideal model for this problem. The increase can be explained by the gambler's ruin population-sizing model [70], which predicts that the population size should grow exponentially with the size of the building blocks.

Figure 4.11 shows the population size and the total number of evaluations until the optimum was found on trap-5 of $n = 100$ to $n = 250$ bits. Figure 4.12 shows the population size and the total number of evaluations until the optimum was found on deceptive-3 of $n = 60$ to $n = 240$ bits.

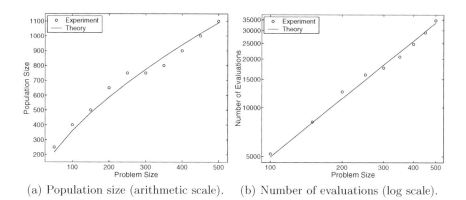

(a) Population size (arithmetic scale).   (b) Number of evaluations (log scale).

Fig. 4.10. BOA performance on onemax of $n = 100$ to $n = 500$ bits. The required population size and the number of evaluations until the optimum was found are shown and compared to the theory. The empirical results match theory well

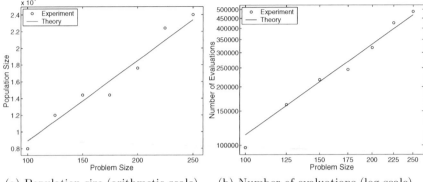

(a) Population size (arithmetic scale).   (b) Number of evaluations (log scale).

Fig. 4.11. BOA performance on trap-5 of $n = 100$ to $n = 250$ bits. The required population size and the number of evaluations until the optimum was found are shown and compared to the theory. The results match the theory well

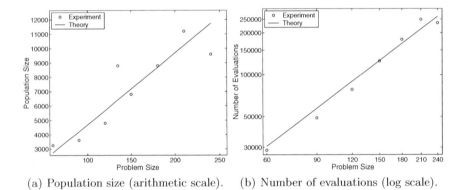

(a) Population size (arithmetic scale).   (b) Number of evaluations (log scale).

Fig. 4.12. BOA performance on deceptive-3 of $n = 60$ to $n = 240$ bits. The required population size and the number of evaluations until the optimum was found are shown and compared to the theory. The empirical results match theory well

The latter two problems (trap-5 and deceptive-3) require that the model learned by BOA is correct. The population-sizing model for building a correct model must therefore be considered. The expected number of evaluations should grow as $O(n^{1.55})$. The results show that the BOA scalability theory approximates the expected number of evaluations well, although as discussed in the previous chapter, a better fit is obtained by $O(n^{1.65})$. We believe that the differences observed are due to the effects of finite populations and the approximations made in developing the theory. Nonetheless, in both cases the empirical results agree with the predicted ones and we can thus conclude that the theory approximates the actual performance well.

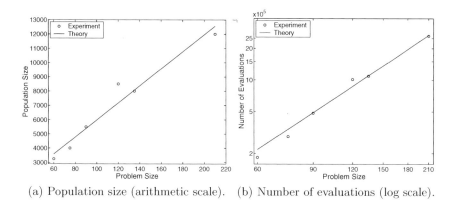

(a) Population size (arithmetic scale).   (b) Number of evaluations (log scale).

Fig. 4.13. BOA performance on exponentially scaled deceptive-3 of $n = 60$ to $n = 240$ bits. The required population size and the number of evaluations until the optimum was found are shown and compared to theory. The empirical results match theory well

4.7.2 Exponential Scaling

To verify the theory developed for exponentially scaled problems, we modified deceptive-3 by scaling the partitions corresponding to the deceptive functions so that the signal decreases exponentially in a specified sequence of partitions. To ensure that a particular partition in the sequence overshadows the contributions of all the subsequent partitions in the sequence, it is sufficient to multiply the ith partition by c^i, where c satisfies the following inequality:

$$0.1c^{\frac{n}{3}} > \frac{n}{3} - 1 . \tag{4.93}$$

The last equation restricts c so that the smallest signal coming from ith partition is greater than the sum of the maximum signals coming from the remaining partitions on positions 1 to $(i-1)$ in the sequence.

Figure 4.13 shows the required population size and the total number of evaluations until the optimum was found for the function described above of $n = 60$ to $n = 210$ bits. Similarly as in the case of uniform scaling, the theoretical growth of the population size and the number of evaluations was compared with the empirical results. The empirical results agree with the predicted ones and we can thus conclude that the theory approximates the actual performance well.

5

The Challenge of Hierarchical Difficulty

Thus far, we have examined the Bayesian optimization algorithm (BOA), empirical results of its application to several problems of bounded difficulty, and the scalability theory supporting those empirical results. It has been shown that BOA can tackle problems that are decomposable into subproblems of bounded order in a scalable manner and that it outperforms local search methods and standard genetic algorithms on difficult decomposable problems. But can BOA be extended beyond problems of bounded difficulty to solve other important classes of problems? What other classes of problems should be considered?

The purpose of this chapter is twofold. First, the chapter poses the challenge of solving problems that are not decomposable into subproblems of bounded order, but can be decomposed over a number of levels of difficulty into a hierarchy. The chapter identifies important features that must be incorporated into BOA to solve such problems in a scalable manner. Second, the chapter presents a class of hierarchically difficult problems that challenge any optimization algorithm because of three inherent features of this class of problems. First of all, the designed problems are not decomposable into subproblems of bounded difficulty. Second, the problems contain exponentially many local optima that make the problems unsuitable for any local search. Finally, the problems deceive any optimization technique based on local operators away from the global optimum.

The chapter starts by motivating the use of hierarchical decomposition in reducing problem complexity. Section 5.2 introduces basic concepts of hierarchical problem solving and presents the important features that must be incorporated into BOA to extend its applicability to hierarchical problems. Section 5.3 proposes the class of problems called hierarchical traps that challenge any problem solver and bound the class of hierarchically difficult problems.

Martin Pelikan: *Hierarchical Bayesian Optimization Algorithm*, StudFuzz **170**, 89–103 (2005)
www.springerlink.com   © Springer-Verlag Berlin Heidelberg 2005

5.1 Hierarchical Decomposition

Many complex systems and processes in business, engineering, science, as well as nature, are hierarchical. By hierarchy, we mean a system composed of subsystems each of which is a hierarchy by itself until we reach some bottom level [171]. Interactions within each subsystem are of much higher magnitude than interactions between the subsystems. There are plenty of hierarchy examples around us. A human body is composed of organs, organs are composed of tissues, tissues are composed of cells, and so on. A university is composed of colleges, colleges are composed of departments, departments are composed of laboratories and offices, and so on. A program code is composed of procedures and functions, procedures are composed of single commands and library calls, commands are composed of machine code or assembly language, and so on.

Why do we talk about hierarchy when what we are really interested in is problem solving and optimization? Most of the complex problems that humans have successfully solved could be tackled only because of the use of hierarchical decomposition. Single-level decomposition discussed in the preceding chapters simplifies a problem by allowing the solver to focus on multiple simpler problems instead of one large problem. However, not every problem can be decomposed into tractable subproblems on a single level; such decomposition may be obstructed due to the rich interaction structure of the problem or the lack of feedback for discriminating alternative solutions to the different subproblems in a fine enough decomposition. Hierarchical decomposition adds a new level of complexity reduction by allowing the decomposition to go down a number of levels until we finally get to a set of subproblems that we can solve. The subproblems in a decomposition on each level are allowed to interact, but the interactions within each subproblem must be of much higher magnitude than those between the subproblems.

5.2 Computer Design, von Neumann, and Three Keys to Hierarchy Success

This section identifies three important concepts that must be incorporated into a selectorecombinative optimizer to solve difficult hierarchical problems quickly, accurately, and reliably. An intuitive example drawn from computer design informally motivates each concept, which is then formalized and discussed in the context of computational optimization and search.

For many of us, computers have become just as important part of our everyday life as cars, refrigerators, and toasters. However, the computer design is an extremely difficult task and many great minds contributed to the effort of making that dream come true. One of the most important contributions to this effort was that of John L. von Neumann who decomposed the design of the "general-purpose computing machine" into four basic components, each of which focused on one task, in particular, arithmetic, memory, control, and

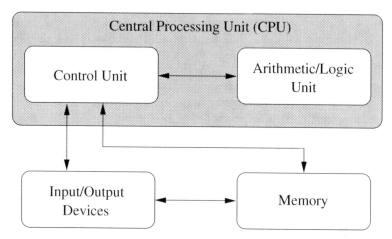

Fig. 5.1. Von Neumann's computer architecture consists of the four main units: (1) control unit, (2) arithmetic/logic unit, (3) input/output unit, and (4) memory unit. The component composed of the control unit and the arithmetic/logic unit forms what we now call the central processing unit (CPU)

connection with the human operator. These four subsystems have formed the basis of most computers up to date as the arithmetic/logic unit, the memory, the control unit, and the input/output devices (see Fig. 5.1).

Further reduction of complexity calls for the use of hierarchy. Each of the four components can be simplified by an additional level of decomposition. For example, a typical hard drive is composed of the controller, heads, magnetic disks, and so forth. Decomposition can go down a number of levels until the problems on the bottom level become tractable.

Although the components on each level in the computer decomposition interact, the design of these components can be oftentimes considered independently. For example, the design of the memory unit does not depend much on the design of the input/output interface. That is why the researchers that are trying to make computer memory chips smaller, faster, and cheaper, do not have to consult each their invention or experience with the ones developing easier-to-use, more powerful, and cheaper input/output devices. Decomposition (and repeated decomposition) is where the simplification of the design comes from. This leads to the first key to hierarchy success:

Key 1: Proper decomposition. On each level of problem solving, the hierarchical black-box optimizer must be capable of decomposing the problem properly. The decomposition allows the solver to focus on the different subtasks in separation and thus reduce complexity. The decomposition can be as simple as the partitioning of the decision variables in the problem into several disjoint subsets but it can be as complex as the graphs with a bounded number of neighbors of each node.

When a particular memory device, such as a hard drive, has been constructed, it becomes irrelevant to consider the details concerning the material or technology used; only several important features suffice. Elimination of irrelevant features of the subsystems from the lower level of the hierarchy reduces the complexity of juxtaposing the subsystems on the current level. In the hierarchical problem solver, it is desirable to eliminate irrelevant details of the solutions from lower levels and represent these in a compact way. This leads to the second key to hierarchy success:

Key 2: Chunking. The solutions to the subproblems from the lower level can be seen as *chunks* of the solutions that are used as basic building blocks for constructing the solutions on the current level. In other words, the partial solutions to each subproblem can be treated as the instances of a single variable that is used to compose the solutions on the next level. The hierarchical problem solver must be capable of representing these chunks of solutions from lower levels in a compact way so that only relevant features are considered.

There are many alternative processors, motherboards, memory chips, and hard drives, which can be used to construct the computer system. The requirements on the final design may suggest which of the alternatives is the best one, but the feedback on the current level may be insufficient to do so. Some of these hard-to-decide choices in the history of computer architecture include the number of machine instructions of the central processing unit (CPU) and the number of registers used by CPU. Furthermore, the features of one component may strongly influence the requirements on another component. For instance, better control devices for hard drives enable faster processing of the requests and allow more complex components ensuring the physical task of storing and retrieving the data. That is why it is necessary that multiple alternatives are maintained for each component and the final choice is made only when there is enough feedback to discard some alternatives. This leads to the third key of hierarchy success:

Key 3: Preservation of alternative candidate solutions. The hierarchical problem solver must be capable of preserving multiple alternative solutions to each subproblem. There are two reasons for doing this:
(1) On the current level there may not be a sufficient feedback to discriminate among a few best alternative solutions to the considered subproblem.
(2) Although the subproblems on the current level are considered independent, interactions on some higher level or levels may lead to new information that favors some of the alternatives over others.

The preceding chapters have shown that BOA is capable of learning an adequate problem decomposition and using the learned decomposition to juxtapose promising partial solutions corresponding to each subproblem. BOA is thus a good starting point in the design of a hierarchical problem solver or optimizer. However, the remaining keys to hierarchy success – chunking and preservation of alternative solutions – must first be incorporated into the basic algorithm. We return to this topic in the following chapter. Here we continue by describing the class of hierarchical problems that challenge any optimization algorithm, because they are not decomposable on a single level and because they are intractable using most popular optimization techniques. Additionally, the designed class of problems should clarify the concept of hierarchical problem solving and decomposition. Finally, the proposed problems bound the class of hierarchically decomposable problems.

5.3 The Design of Challenging Hierarchical Problems

It is common in the design of material machines like airplanes and toasters to test the design at the boundary of its design envelope. To test an algorithm on whether it is capable of exploiting decompositional bias in a scalable manner, additively separable traps introduced earlier can be used. However, all the previously discussed problems were solvable by a single-level decomposition without forcing the use of a more robust and general bias based on hierarchy.

The purpose of this section is to design a class of challenging hierarchical problems that can be used to test the scalability of optimization algorithms on difficult hierarchical problems. The design is guided by the three keys to hierarchy success presented in the previous section: (1) proper decomposition, (2) chunking, and (3) preservation of alternative solutions. The section starts by introducing a general class of hierarchically decomposable test problems. Next, the section motivates the design of the class of *hierarchical trap functions* and presents several such functions.

5.3.1 Example: Tobacco Road

To get a better idea of what we mean by a hierarchical function – in particular, a challenging hierarchical function – let us start with an example called the *tobacco road function* introduced by Goldberg [55]. To better understand the example, please follow Fig. 5.2.

The tobacco road function is defined on two levels. On the bottom level, the input string is partitioned into partitions of 6 bits each, where each partition is evaluated using the folded trap function of order 6 (see Fig. 5.3). The partitioning can be chosen arbitrarily but it remains fixed for all evaluations. The folded trap has two *global* optima: 000000 and 111111. Additionally, the folded trap function contains a large number of local optima in all the strings that contain half 1s and half 0s, which is right in the middle between the two global optima. Similarly as the 5-bit trap, the folded trap cannot be decomposed without losing the global optima and converging to some of the local ones. Moreover, the deception in the folded trap is somewhat more harmful

1. Fitness evaluation on the 1st level.

$$f = 1 + 0.9 + 0.9 + 1 + 1 + 1 = 5.8$$

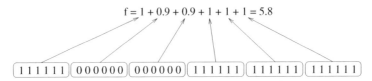

2. Mapping the partitions to the 2nd level.

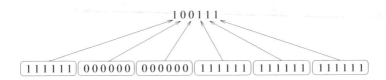

3. Fitness evaluation on the 2nd level.

$$g = 1.0$$

100111

4. Computation of the overall fitness.

$$fitness = f + g = 5.8 + 1.0 = 6.8$$

Fig. 5.2. The tobacco road function is defined on two levels. The solution on the bottom level is first evaluated using folded traps and then mapped to form the solution on the second level. The second-level solution is then evaluated using traps of order 6 and the overall fitness is computed by adding the contributions of both levels together

than the one in the single trap, since the folded trap contains many more local optima than the global ones.

The bits in each partition corresponding to one of the folded traps are then mapped to one symbol on the next level: 000000 maps to 0, 111111 maps to 1, and the rest maps to a null value that stands for an undefined bit and is denoted by '-'. After mapping the partitions, the next level contains six times less bits than the bottom level and some of those bits may be undefined.

The new symbols on the second level are then partitioned into partitions of 6 bits. Again, the partitioning is fixed for all evaluations. Note that each partition of 6 bits on the second level is determined by some subset of 36 bits on the bottom level. Each of the 6-bit partitions of the second level contributes to the overall fitness function using a trap function of order 6, defined analogously to the 5-bit trap:

$$trap_6(u) = \begin{cases} 6 & \text{if } u = 6 \\ 5 - u & \text{otherwise} \end{cases}, \tag{5.1}$$

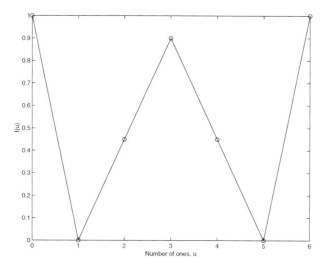

Fig. 5.3. The folded trap function contains two global optima, one in the string of all 0s, and one in the string of all 1s. Additionally, the folded trap contains a number of deceptive attractors in the strings with half 0s and half 1s. Deception and multimodality make folded traps a great challenge for any optimization method

where u is the number of ones in the input string of 6 symbols. However, if a partition contains at least one null symbol, the contribution of the partition is set to 0.

The overall fitness is computed by adding the fitness contributions on both levels. The global optimum of a tobacco road function is in the string of all 1s. Tobacco roads also contain a large number of local optima. A solution is a local optimum if it sets each trap partition on the first level to contain either 0, 3, or 6 copies of a 1, but if at least one of these partitions is not 111111.

The tobacco road function can be solved in either one or two stages. If the algorithm approaches the tobacco road in one stage, the decomposition must consider building blocks of 36 bits. Due to the factor of 2^k in the decision-making (for simple GAs) or model-building (for BOA) population-sizing model, this approach is highly inefficient. Nonetheless, considering sub-problems of such a high order is unnecessary. Solving the problem in two stages yields much more promising results. In particular, the first phase can focus on optimizing the bottom level yielding locally optimal solutions with blocks 000000 and 111111 on all partitions of the bottom-level decomposition. If the algorithm was capable of recognizing that blocks 000000 and 111111 actually represent the two most promising alternatives and treated them as if they were single bits, the second stage would consist of solving an additively separable problem consisting of 6-bit traps. In this manner, blocks 111111 would be combined to form the optimum.

The difference between the two approaches – the single-stage, single-level approach and the two-stage, hierarchical approach – is in the order of subproblems that must be considered to find the optimum. An important observation is that the number of fitness evaluations will differ by a factor of about 2^{30}, because the overall complexity grows exponentially with the order of the problem decomposition. In other words, just considering the problem on two levels decreases the number of evaluations by a factor of more than one billion – more precisely, $1,073,741,824$.

The tobacco road function is also extremely difficult for any local search. Since the blocks of 36 bits must be considered if approached on a single level, the performance of the hill climber would grow with $O(n^{36} \ln n)$. Although n^{36} is a polynomial, solving the problem in $O(n^{36} \ln n)$ evaluations becomes intractable incredibly fast.

Therefore, there is no question whether the tobacco road functions are challenging for BOA or any other optimization algorithm. Furthermore, the difference between the hierarchical and single-level approaches increases with the number of levels and even more difficult problems could be therefore designed in a straightforward manner.

The following section formalizes the notion of hierarchically decomposable problems. Subsequently, the class of hierarchical trap functions is introduced where the number of levels grows with the problem size.

5.3.2 Hierarchically Decomposable Functions

The idea of using hierarchical functions in challenging GAs and other optimization techniques dates back to the works on royal road functions of Mitchell et al. [109], hierarchical if-and-only-if functions of Watson et al. [181], tobacco road functions of Goldberg [55], hyperplane-defined functions of Holland [84], and others [126, 127]. This section attempts to formalize all of the aforementioned approaches similarly as it was done by Watson et al. [181], and Pelikan and Goldberg [126].

Hierarchically decomposable functions (HDFs) are defined on multiple levels. The solution on each level is partitioned according to the structure of the function. The bits can be partitioned arbitrarily, but the partitioning must be fixed for all evaluations. Partial solutions in each partition are evaluated according to a given function. Each partition is then mapped to the next level into a single symbol. The mapping functions can be chosen arbitrarily. The overall fitness is computed by summing the contributions of all the partitions on all the levels.

More formally, an HDF can be defined by the following three components (see Fig. 5.4 for the definition of the components for the tobacco road function):

1. **Structure.** The structure of the HDF defines the way in which the variables on each level are partitioned. Partitions are used to both contribute

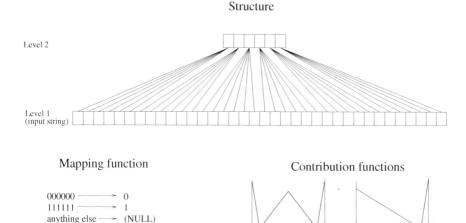

Fig. 5.4. The three components defining a 36-bit tobacco road function

to the fitness of the solution and map to the next level. The structure forms a tree with one-to-one mapping between the leaves of the tree and the problem variables (string positions). Here we only use balanced trees of fixed order. The levels are numbered from the bottom – the first level consists of the leaves, whereas the last level consists of the children of the root or the root itself. In most cases we omit the root.

2. **Mapping functions.** The mapping functions, also called the interpretation functions, define how to map solutions from a lower level to become inputs of both the contribution and interpretation functions on the current level.

3. **Contribution functions.** The contribution functions define how much the partitions of the decomposition on each level contribute to the overall fitness. Since the mapping of the solution to the next level reduces the number of partitions on the next level, it is often useful to scale the contributions on each level so that the total contribution of each level remains the same.

Before describing hierarchical traps, we present the aforementioned functions proposed in the past that contain some form of hierarchy and discuss their problem difficulty. Later we shall argue that hierarchical traps challenge optimization methods in a qualitatively different way and that without "cheating" there is no way of getting around the difficulty introduced by hierarchical traps.

5.3.3 Another Example: Royal Road

The class of royal road functions was proposed by Mitchell et al. [109] as a way of describing the hierarchical nature of optimization in GAs and providing an example function that GAs were born to solve. Although royal roads lack in several aspects of problem difficulty that we are interested in, the idea of hierarchy in GA problem solving manifested by GA behavior on royal roads is an interesting and influential one.

A royal road function is defined by the list of fitness contributions of a subset of partial solutions [109]. Here we only discuss a simple royal road function adopted from Mitchell et al. [109] that contains hierarchy and thus relates to the topic of hierarchical difficulty studied in this chapter.

The fitness of the royal road function studied here is computed in several stages. In the first stage, the input binary string is partitioned into partitions of 8 consecutive bits each. Each block of bits in these partitions contributes to the overall fitness by 8 if the block is 11111111; otherwise, the block does not contribute to the overall fitness at all. Each subsequent stage merges the pairs of neighboring blocks together, yielding half as many blocks of twice as large order. In the second stage, for instance, blocks of 16 consequent bits each are created. Each of the created blocks of bits then contributes to the overall fitness by the number of bits included in the block if all the bits in the block are 1. So in the second stage, for instance, each block 1111111111111111 contributes to the overall fitness by 16. The evaluation continues until the block covers the entire string; the string length should therefore be a power of 2 multiplied by 8. See Fig. 5.5 for the three-component HDF definition of the royal road.

The royal road described above has one global optimum in the string of all 1s and no local optima. The fitness contributions of the blocks of 1s grow with the size of the blocks and at any stage of the search, the signal leads

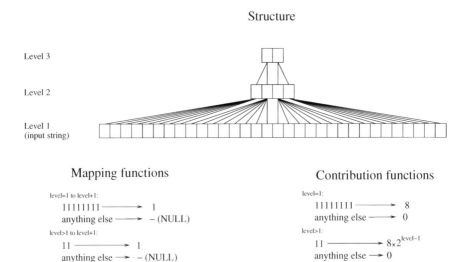

Fig. 5.5. The three components defining a 32-bit royal road function

to the global optimum. Consequently, GAs are able to solve the royal roads efficiently. However, because of the same reasons, the royal roads are easy even for the stochastic hill climber described earlier. The reason for this is that in the royal roads, there is nothing to mislead the hill climber away from the global optimum and it is sufficient that the hill climber is capable of crossing 8-bit plateaus. Even more importantly, the royal roads do not challenge an optimization algorithm to obey the three keys to hierarchy success and are efficiently solvable by any standard GA. Consequently, if the algorithm provides scalable solution for royal roads, one cannot imply that it also provides scalable solution for arbitrary decomposable problems of bounded difficulty or hierarchically decomposable problems.

5.3.4 Yet Another Example: Hierarchical if-and-only-if (HIFF)

The hierarchical if-and-only-if function was proposed by Watson et al. [181], Watson [180] as an example of a function that is not separable and should therefore challenge even those GAs that are capable of finding building blocks of bounded order.

The structure of HIFF is a balanced binary tree. The input to the contribution and mapping functions therefore consists of two symbols. A single mapping function is used on all levels where 00 is mapped into 0, 11 is mapped into 1, and everything else is mapped into the null symbol '-'. On each level, blocks 00 and 11 contribute to the overall fitness by 2^{level}, where $level$ is the number of the current level (again, the bottom level is level 1). Anything else does not contribute to the overall fitness. Additionally, each leaf in the tree (a single bit) contributes to the fitness by 1. Since the structure is a balanced binary tree, the size of the problem should be a power of 2. Figure 5.6 shows the three HDF components defining HIFF.

HIFF has two global optima, one in the string of all 1s and one in the string of all 0s. To successfully solve HIFF, an optimizer must preserve either 0s or 1s on all string positions to ensure that the optimum can be reached. Although the loss of particular bits on some positions can be taken care of by mutation or another local operator, the complexity of changing any partition grows exponentially with its order and, therefore, there must be a limited number of such problematic positions. There are two alternative ways of solving HIFF. The algorithm can decide whether to go after 0s or 1s, or preserve both alternatives as the optimization proceeds. In the latter case, the algorithm must ensure preservation of the partitions on the current level of optimization, because mixing 0s with 1s moves the optimization one or more levels down. Of course, the chunks of 0s and 1s must be combined together efficiently.

The following section presents hierarchical traps, which add a new source of difficulty to tobacco road functions and HIFF by combining the ideas of the two designs into one.

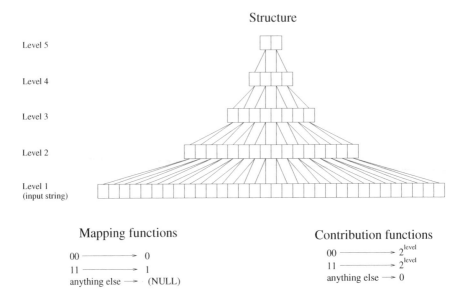

Fig. 5.6. The three components defining a 32-bit HIFF function

5.3.5 Hierarchical Traps: The Ultimate Challenge

Before designing hierarchical traps, let us recall the important goals of the design:

1. **Force proper decomposition.** Hierarchical traps must force the algorithm to learn an appropriate problem decomposition on each level. In other words, if the algorithm fails to find an adequate decomposition on any level, computational complexity should grow with problem size prohibitively fast or the optimum should not be found at all.
2. **Force preservation of alternative solutions.** Hierarchical traps must force the algorithm to preserve alternative partial solutions. Failing to do this should result in either intractable or unsuccessful search for the global optimum.
3. **Force chunking.** Hierarchical traps must force the algorithm to manipulate large pieces of solutions. Similarly as in the above two cases, failing to do this should result in either intractable or unsuccessful search for the global optimum.

To ensure the achievement of all the above goals, hierarchical traps combine the features of the tobacco road function and HIFF. We start by presenting the general definition of hierarchical traps. Subsequently, we specialize the general definition yielding two challenging hierarchical trap problems, f_{htrap1} and f_{htrap2} that are used in the remainder of this book.

The three components defining a general hierarchical trap are listed in the following list (see also Fig. 5.7):

Structure

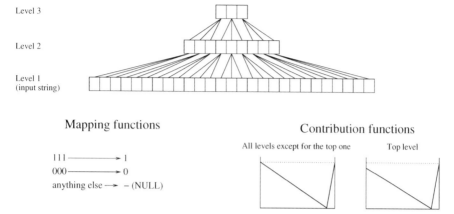

Fig. 5.7. The three components defining a 27-bit hierarchical trap of order $k = 3$ consisting of 3 levels

1. **Structure.** Hierarchical traps use a balanced k-ary tree as the underlying structure, where $k \geq 3$. The lower bound on k is given by the minimum order of fully deceptive functions that are to be used as contribution functions. Since we are interested in the scalability of tested optimization algorithms and the problem sizes must grow as powers of k, it is reasonable to set k to a small value, preferably 3.
2. **Mapping Functions.** One mapping function is used to map all partitions on all levels. The mapping function maps blocks of all 0's and 1's to 0 and 1, respectively, similarly as in the tobacco road and HIFF. Everything else is mapped to '-'.
3. **Contribution Functions.** Contribution functions are based on traps of order k (see Fig. 1.5 for an example trap function of order 5):

$$trap_k(u) = \begin{cases} f_{high} & \text{if } u = k \\ f_{low} - u\frac{f_{low}}{k-1} & \text{otherwise} \end{cases} . \tag{5.2}$$

The height of the two optima in each trap (denoted by f_{low} and f_{high}) depends on the current level. The difficulty of hierarchical traps can be tuned by parameters f_{low} and f_{high} on different levels.

Let us now specify the parameters for the two hierarchical traps that are going to be used in our experiments, denoted by f_{htrap1} and f_{htrap2}, respectively. In both f_{htrap1} and f_{htrap2}, the underlying structure is a ternary tree (i.e., $k = 3$). The two hierarchical traps differ in the parameters f_{low} and f_{high} on each level.

The hierarchical trap f_{htrap1} sets $f_{high} = f_{low} = 1$ on all levels except for the top one. That means that in the traps on all the levels except for the top one, the two optima are equally good. In the trap on the top level, the optimum in $u = 3$ is $f_{high} = 1$, whereas the optimum in $u = 0$ is $f_{low} = 0.9$. The contributions on each level are multiplied by 3^{level} so that the total contribution of each level is the same. It is easy to see that the f_{htrap1} has one global optimum in the string of all 1s. Additionally, f_{htrap1} has a large number of local optima. There are several features of f_{htrap1} worth mentioning at this point:

- There is no way of deciding whether 000 is better than 111 in any subproblem on any level except for the top one. Furthermore, a correct decision between 000 and 111 can be made only if the global optimum has already been found. For example, the solution 111111000 has the same fitness as the solution 000111000, although the latter solution is much farther from the global optimum in 111111111.
- For each subproblem on any level, 000 is easier to find than 111 because it is surrounded by high-quality solutions. Therefore, we can expect that the number of 1s decreases as the search goes up a number of levels.
- Because of the character of traps, small, local changes in non-optimal solutions lead *away from the optimum* in most cases.

The hierarchical trap f_{htrap2} also uses the trap with $f_{high} = 1$ and $f_{low} = 0.9$ on the top level. However, f_{htrap2} makes the bias toward the solutions containing many 0s even stronger by making the peak f_{low} higher than the peak f_{high} on all lower levels (all levels except for the top-most level). Therefore, on all lower levels, the optimum in 111 is not only isolated but also local. To ensure that the optimum still remains in the string of all 1s, values of f_{low} and f_{high} used on lower levels must satisfy the following inequality:

$$(l - 1)(f_{low} - f_{high}) < 0.1 , \qquad (5.3)$$

where l is the total number of levels. The last equation must be satisfied so that the difference between the peaks on the top level is greater than the sum of the differences on the remaining levels. We set $f_{high} = 1$ and $f_{low} = 1 + 0.1/l$ on all levels except for the top one. Similarly as in the previous case, the contributions on each level are multiplied by 3^{level}. There is one additional difficulty of searching f_{htrap2} compared to f_{htrap1}:

- If the algorithm does not preserve 111 despite that 000 performs slightly better, the optimum cannot be found. Recall the example presented earlier in this section. For f_{htrap2}, despite that 111111000 is much closer to the global optimum in 111111111 than 000111000 is, the fitness of 000111000 is greater than the fitness of 111111000 (while in f_{htrap1} the fitness is the same for both solutions).

Hierarchical traps presented above differ from the previously proposed hierarchical functions in several aspects. The path to the global optimum in hierarchical traps is much windier than the one in the royal roads. Not only are the solutions with many 0s easier to find, but if these are disrupted the search must go down one or more levels. Unlike HIFF, hierarchical traps do not allow the algorithm to decide whether to go toward 0s or 1s; the global optimum is in the string of all 1s, although for all levels except for the top one it is either impossible to distinguish between 0s and 1s, or 0s look even better than 1s. The bias toward solutions with many 0s distinguishes hierarchical traps from HIFF and introduces a bigger challenge. Hierarchical traps and tobacco roads share many similarities. However, hierarchical traps contain more levels than tobacco roads do and by using low-order subproblems they make the scalability analysis more practical.

6

Hierarchical Bayesian Optimization Algorithm

The previous chapter has discussed how hierarchy can be used to reduce problem complexity in black-box optimization. Additionally, the chapter has identified the three important concepts that must be incorporated into black-box optimization methods based on selection and recombination to provide scalable solution for difficult hierarchical problems. Finally, the chapter proposed a number of artificial problems that can be used to test scalability of optimization methods that attempt to exploit hierarchy.

The purpose of this chapter is to extend BOA to solve difficult hierarchical problems quickly, accurately, and reliably. The chapter discusses several such extensions and implements one of them. The proposed algorithm is called the hierarchical BOA (hBOA). hBOA is then tested on the challenging hierarchical problems presented in the previous chapter.

The chapter starts by discussing the first two keys to hierarchy success – proper decomposition and chunking. Section 6.1 describes one of the approaches to tackling both keys at once. More specifically, the section incorporates local structures into Bayesian networks to allow compact representation of the local conditional distribution of each variable. Section 6.2 focuses on the last key to hierarchy success – preservation of alternative solutions – and reviews approaches to maintaining useful diversity in genetic and evolutionary computation. The section classifies presented diversity-maintenance techniques into several categories and discusses them in the context of BOA. Section 6.3 summarizes the extensions of the original BOA that comprise hBOA. Section 6.4 analyzes hBOA performance on several hierarchical problems, including hierarchical traps and HIFF. Finally, Sect. 6.5 discusses hBOA performance in the context of the BOA scalability theory presented in Chap. 4.

6.1 Proper Decomposition and Chunking

Chapters 3 and 4 have shown that BOA is capable of learning and exploiting decomposition of problems that are decomposable into subproblems of

Martin Pelikan: *Hierarchical Bayesian Optimization Algorithm*, StudFuzz **170**, 105–129 (2005)
www.springerlink.com   © Springer-Verlag Berlin Heidelberg 2005

a bounded order on a single level. Learning and exploitation of an adequate problem decomposition is ensured by (1) constructing a Bayesian network that captures important nonlinearities in the set of promising solutions and (2) sampling the learned network to generate new candidate solutions. Bayesian networks can also be used to encode, learn, and utilize decomposition of hierarchical problems on any level. However, additionally to representing dependencies and independencies in the problem on the current level, the graphical model must incorporate some form of chunking as described in Sect. 5.2.

The purpose of this section is to review alternative approaches to incorporating chunking into BOA without compromising BOA's capability of decomposing the problem properly. Since the issues of chunking and learning a proper decomposition are strongly correlated, the two issues are addressed together.

The section first discusses different facets of chunking in detail. The section then describes how local structures – such as default tables, decision trees, and decision graphs – can be used to enhance Bayesian networks and, in turn, ensure chunking.

6.1.1 Chunking Revisited

The primary goal of chunking is to allow groups of variables from each subproblem of the lower level to be merged into a single variable (or an intact block) that encodes all the relevant information needed to distinguish alternative partial solutions to this subproblem. These merged variables serve as the basic building blocks for juxtaposing solutions on the next level. Note that there are two problems that must be considered:

1. **Merging.** The model must be capable of merging a group of variables from one subproblem on the lower level into a single variable or a block. This can be done either explicitly or implicitly, but the model must allow such merging to take place.
2. **Representing partial solutions efficiently.** The model must represent partial solutions compactly so that only relevant information is considered. Compact representation is necessary for ensuring that large partial solutions can be encoded in the model (either as an intact block or as a new variable).

Merging the variables into groups can be incorporated by creating specialized models that can encode dependencies and independencies among the groups of variables as opposed to dependencies and independencies among the variables themselves. Nonetheless, since Bayesian networks have been shown to be capable of representing the chunks of solutions of decomposable problems in BOA, the issue of merging the variables can be taken care of by using traditional Bayesian networks. But in either case, efficient representation of the relevant features of each chunk must be ensured.

There are two approaches to incorporating chunking into BOA:

1. **Explicit chunking.** Modify the model so that partitions of solution strings are allowed to form a single component of the model. The relationships between the chunks can then be represented by a Bayesian network like in BOA or in some other way. Example models that allow explicit chunking are the marginal product models (MPM) of ECGA [70], Huffman networks that combine MPM with Bayesian networks [30], and Bayesian networks with hidden variables [27, 50]. To ensure explicit chunking, each of these models must represent its parameters efficiently so that large groups of variables can be processed.
2. **Implicit chunking.** Ensure that the used model is capable of representing the chunks, although the overall structure might not correspond to an intuitive notion of chunking based on merging the variables into groups explicitly. Additionally, each model must represent its parameters efficiently so that large chunks can be represented. Bayesian networks with local structures, discussed in the following section, represent one class of models capable of implicit chunking.

Since in both approaches, compact representation of the model is necessary and Bayesian networks can encode chunks at no extra cost, the remainder of this section focuses on implicit chunking. In particular, the section describes local structures that can be used in Bayesian networks to encode high-order dependencies efficiently.

6.1.2 Local Structures in Bayesian Networks

The number of conditional probabilities that must be specified in a Bayesian network grows exponentially with the order of interactions encoded by the network. Although in some cases, simple models suffice and there is no need for compact representation, exponential growth of the number of model parameters can become a bottleneck on more complex problems.

This section describes how local structures can be used to make representation of conditional distributions in Bayesian networks more compact and to allow a probabilistic model to encode high-order interactions. Additionally to providing a technique for chunking, local structures allow learning and sampling more complex models with a reasonable number of parameters.

The section starts by providing an example conditional probability table that motivates the use of local structures. Next, the section describes default tables that specify only a subset of the conditional probabilities and set the remaining probabilities to a specified constant. Finally, the section introduces decision trees and graphs that reduce the number of parameters in a more general fashion.

Motivating Example

To fully specify the local probability distribution of a binary variable conditioned on k other binary variables, a conditional probability table with 2^k probability entries can be used. Note that there are 2^{k+1} combinations of values of the variable and its parents, but conditional probabilities for half of them can be computed using the following rule relating probabilities of complementary events:

$$p(X_i = 0|\Pi_i) = 1 - p(X_i = 1|\Pi_i) , \qquad (6.1)$$

where X_i denotes the considered variable and Π_i denotes the parents of X_i.

To motivate the use of local structures in Bayesian networks, let us consider a binary variable X_1, which is conditioned on 3 other binary variables denoted by X_2, X_3, and X_4. Therefore, the full conditional probability table for X_1 contains 8 entries. In our example, the probabilities are set as shown in Table 6.1.

Table 6.1. A conditional probability table for X_1 whose parents are X_2, X_3, and X_4. Only the probabilities of one of the values of X_1 must be stored, because the remaining probabilities can be computed using $p(X_1 = 1|X_2, X_3, X_4) = 1 - p(X_1 = 0|X_2, X_3, X_4)$

X_2	X_3	X_4	$p(X_1 = 0|X_2, X_3, X_4)$
0	0	0	0.75
0	0	1	0.25
0	1	0	0.25
0	1	1	0.25
1	0	0	0.20
1	0	1	0.20
1	1	0	0.20
1	1	1	0.20

Although 3 entries in Table 6.1 are 0.25 and 4 entries are 0.20, the conditional probability table lists each single probability entry. Listing all probabilities is easy for dependencies of order 3, but it will become a major bottleneck if we want to represent dependencies of order 50, for instance ($2^{50} > 10^{15}$). Local structures allow the use of regularities in the conditional probability tables and encode the probabilities in a more compact way.

The remainder of this section describes three types of local structures – default tables, decision trees, and decision graphs – that can be used to encode conditional probabilities in a Bayesian network. Of course, there are many other ways to approach the same problem, but the approaches presented in this section should provide a sufficiently powerful means for ensuring compact representation of Bayesian networks. The conditional probability table shown in Fig. 6.1 will be used throughout this section to illustrate the effectiveness of different types of local structures and their basic principle.

6.1.3 Default Tables

A default table lists several probabilities with the corresponding instances of the variables in the condition first, and sets the remaining probabilities to a default value. The default value is computed as a weighted average of all covered conditional probabilities where each probability is weighted using the probability of its condition. If a particular conditional probability is listed, its value is determined from the corresponding entry in the table. If the conditional probability is not listed, its value is assumed to be equal to the default value.

There are two intuitive ways to compress the conditional probability table from Table 6.1 using default tables. The first approach is to list the first four probabilities and set the remaining probabilities to the default value of 0.20. The second alternative is to list the first probability and the last four ones first, and set the remaining probabilities to the default value of 0.25. In the first case, more entries in the table are eliminated and therefore higher compression is achieved. The resulting default table is shown in Table 6.2.

Table 6.2. A default table reducing the number of conditional probabilities from Table 6.1 by 3 without compromising the accuracy of the model

X_2	X_3	X_4	$p(X_1 = 0|X_2, X_3, X_4)$
0	0	0	0.75
0	0	1	0.25
0	1	0	0.25
0	1	1	0.25
	default		0.20

The default table presented above reduces the number of probability entries without losing any information at all. However, it is possible to group the probabilities that are not equal but that are very similar. For example, the conditional probability table shown in Table 6.1 can be further reduced as shown in Table 6.3 by merging the probabilities 0.20 and 0.25 into a single default value. Although such an assignment causes the probabilities encoded in the network to differ from those observed in the data, the assignment might still lead to an increase in the score of the model due to the decreased model complexity.

The limitations of default tables are clear; a default table is capable of encoding similarity among the members of only *one* subset of probabilities. That is why one must choose whether to group the probabilities 0.20 or 0.25,

Table 6.3. A default table reducing the number of conditional probabilities by 6. In this case, the conditional probabilities of 0.20 and 0.25 are both stored in the default row of the table, yielding a different assignment of the probabilities than the one observed in the data. However, the default table may still increase the quality of the model with respect to the used scoring metric

X_2	X_3	X_4	$p(X_1 = 0 \vert X_2, X_3, X_4)$
0	0	0	0.75
	default		0.23

but it is impossible to group all the instances with the probability 0.20 and, at the same time, group all the instances with the probability 0.25. Next, we describe how more sophisticated structures – such as decision trees and decision graphs – can be used to exploit more complex regularities in conditional probability tables.

6.1.4 Decision Trees

A decision tree is a directed acyclic graph where each node except for the root has exactly one parent; the root has no parents. Every internal node (a node that is a parent of some other node) of the tree is labeled with a variable. When a node is labeled with a variable v, we say that the node is a split on v. Edges from a split on v are labeled with non-empty disjoint exhaustive subsets of possible values of v. There are at least two edges starting in any internal node.

Given an assignment of all the variables (a candidate solution), a traversal of the decision tree starts in the root. On each split on v, the traversal continues to the child along the edge containing the current value of v. For each assignment of the involved variables, there is only one possible way of traversing the tree to a leaf, because the edges coming from each split must be labeled with disjoint subsets of the values.

Each leaf of a decision tree contains a quantity or information of interest associated with all the instances that end up the traversal of the tree in this leaf. To use a decision tree to represent the conditional probabilities of a variable, each leaf stores the conditional probabilities of this variable given a condition specified by the variable assignments along the path to this leaf.

Let us return to the example from Table 6.1 discussed in the context of default tables. To represent the full conditional distribution, it is sufficient to construct the tree shown in Fig. 6.1(a). The decision tree reduces the number of conditional probability entries that must be stored from 8 to only 4.

In some cases, it might be useful to store fewer probability entries at the expense of the accuracy of the resulting model. Analogously to the case with default tables, the reduction of the number of model parameters might result in increasing the overall model quality.

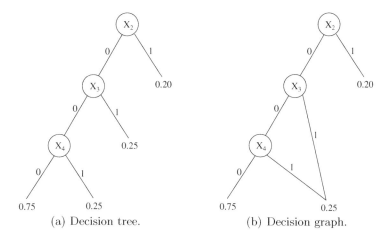

Fig. 6.1. A decision tree and a decision graph that encode the probabilities shown in Table 6.1

6.1.5 Decision Graphs

Note that in the decision tree from Fig. 6.1(a), there are two leaves that store the same conditional probability of 0.25. If the two leaves could be merged into a single one, the number of probability entries could be further reduced. However, after merging the two leaves, the new node would have two parents instead of one; consequently, the resulting graph would not be a tree. In spite of that, the resulting structure would be still useful in the same way.

Decision graphs extend decision trees by allowing each node to have multiple parents and therefore can find a use for more regularities in the conditional probabilities associated with the Bayesian network. For example, the decision graph encoding the probabilities from the aforementioned example is shown in Fig. 6.1(b). Using a decision graph reduces the number of conditional probabilities that must be stored by more than a half (only 3 out of 8 probabilities must be stored).

Parsing a decision graph is analogous to parsing a decision tree. For each instance, there is exactly one way of traversing the graph down to a particular leaf, which stores the corresponding probability. Unlike decision trees, decision graphs allow encoding any equality constraints on the conditional probabilities. In other words, there can be as few leaves in the decision graph as there are different conditional probabilities. By introducing additional equality constraints, the overall quality of the model can be further increased.

6.1.6 Bayesian Network with Decision Graphs

A Bayesian network with decision graphs contains one decision graph for each variable. The decision graph for X_i is denoted by G_i. Each decision graph G_i encodes the conditional probabilities for X_i.

Note that although decision graphs encode conditional probabilities for a network structure, the network structure can be constructed from the decision graphs themselves. The set of parents of X_i is the set of variables on which there exists a split in the decision graph G_i. As long as the decision graphs contain valid probabilities and there are no cycles in the corresponding Bayesian network, the network is fully specified by the set of decision graphs, one graph per variable.

As an example, consider the decision graph shown in Fig. 6.1(b), which stores the conditional probabilities of X_1 shown earlier in Table 6.1. The decision graph reveals that the probability distribution of X_1 is affected by the variables X_2, X_3, and X_4, and it is independent of any other variable. Therefore, X_2, X_3, and X_4 are the parents of X_1 in the underlying network structure.

There are three primary advantages of using decision graphs in learning Bayesian networks:

(1) **Parameter compression.** Fewer conditional probabilities can be used to represent the model. This saves memory and time for the construction, storage, and utilization of the model.

(2) **More complex models.** Decision graphs allow learning a more complex class of models. Although any probability distribution can be represented by a full joint probability distribution represented by a fully connected Bayesian network, the full joint probability distribution requires an exponential number of parameters. Decision graphs can reduce the number of parameters so that learning the model becomes tractable even when there are no valid conditional independencies.

(3) **Better learning.** The construction of a Bayesian network with decision graphs can proceed in smaller and more specific steps. That often results in better models with respect to their likelihood [23, 44].

Using decision graphs improves the utility of Bayesian networks, but the semantics of Bayesian networks with decision graphs remains the same. Therefore, one could expect that although the expressiveness of Bayesian networks will improve by using decision graphs, the methods for measuring quality of candidate models and the methods for constructing a model from data should not change much. Indeed, the scoring metrics for measuring the quality of each candidate model as well as the greedy algorithm for constructing the model can be adapted to the new situation in a straightforward manner.

In the following, we first describe changes that must be incorporated into the two metrics presented earlier – the Bayesian-Dirichlet metric (BD) and the Bayesian information criterion (BIC). Subsequently, we define the split and merge operators, which can be used to construct decision graphs. Finally, we present a greedy algorithm for learning Bayesian networks with decision graphs. The presented algorithm exploits the new representation of Bayesian networks by directly manipulating decision graphs instead of updating the decision graphs as a consequence of changing the network itself.

6.1.7 Bayesian Score for Networks with Decision Graphs

Chickering et al. [23] derived the Bayesian-Dirichlet metric for Bayesian networks with decision graphs. The resulting metric can be computed analogously to the case with traditional Bayesian networks, yielding

$$p(D|B) = \prod_{i=1}^{n} \prod_{l \in L_i} \frac{\Gamma(m_i'(l))}{\Gamma(m_i(l) + m_i'(l))} \prod_{x_i} \frac{\Gamma(m_i(x_i, l) + m_i'(x_i, l))}{\Gamma(m_i'(x_i, l))} , \qquad (6.2)$$

where L_i is the set of leaves in the decision graph G_i for X_i; $m_i(l)$ is the number of instances in D which end up the traversal through the graph G_i in the leaf l; $m_i(x_i, l)$ is the number of instances that have $X_i = x_i$ and end up the traversal of the graph G_i in the leaf l; $m_i'(l)$ represents the prior knowledge about the value of $m_i(i, l)$; and $m_i'(x_i, l)$ represents the prior knowledge about the value of $m_i(x_i, l)$. Again, an uninformative prior $m_i'(x_i, l) = 1$ is used in the K2 variant of the BD metric for Bayesian networks with decision graphs.

As mentioned earlier, Bayesian metrics tend to be more sensitive to the noise in data and, in practice, they often lead to overly complex models. However, Bayesian metrics allow the use of prior information, which can bias the metric to favor simpler models and alleviate the aforementioned problem. Although in traditional Bayesian networks we have not succeeded in finding a robust prior that would work well in general, in Bayesian networks with decision graphs we have resolved this problem. This robust prior is described next.

To adjust the prior probability of each network according to its complexity, the description length of the parameters required by the network is first computed. One frequency in the data set of size N can be encoded using $\log_2 N$ bits; however, only half of the bits suffice to encode the frequencies with sufficient accuracy [45]. Therefore, to encode all the parameters, $0.5(\sum_i |L_i|) \log_2 N$ bits are needed, where $\sum_i |L_i|$ is the total number of leaves in all the decision graphs. To favor simpler networks to the more complex ones, the prior probability of a network can decrease exponentially with the description length of the set of parameters they require [44]. Thus,

$$p(B) = c2^{-0.5(\sum_i |L_i|) \log_2 N} , \qquad (6.3)$$

where c is a normalization constant required for the prior probabilities of all possible network structures to sum to 1. The normalization constant does not affect the result because we are interested in only relative quality of networks and not the absolute value of their marginal likelihood.

Based on our experience, the above prior is sufficient to bias the model construction to networks with fewer parameters and avoid superfluously complex network structures. For local structures in Bayesian networks, the K2 score with the above prior actually seems to outperform the BIC metric in its robustness.

A similar prior was proposed by Chickering et al. [23], who decrease the prior probability of each network exponentially with the number of parameters of the network:

$$p(B) = c\kappa^{\sum_i |L_i|} ,$$

where $\kappa \in (0,1)$. Usually, κ is rather small, for instance, $\kappa = 0.1$.

6.1.8 BIC for Bayesian Networks with Decision Graphs

The Bayesian information criterion (BIC) for Bayesian networks with decision graphs can also be computed analogously to the case with traditional Bayesian networks (see Equation (3.4)). The difference is that instead of summing over the instances of each variable and its parents, the sums must go over all leaves in the tree. Additionally, the number of parameters is not anymore proportional to the number of instances of the parents of all variables, but it is equal to the number of leaves. Thus,

$$BIC(B) = \sum_{i=1}^{n} \left(N \sum_{l \in L_i} m_i(x_i, l) \log_2 \frac{m_i(x_i, l)}{m_i(l)} - |L_i| \frac{\log_2(N)}{2} \right) . \qquad (6.4)$$

6.1.9 Decision Graph Construction: Operators on Decision Graphs

To construct a decision graph for binary variables, two operators are sufficient. The first operator is a *split*, which splits a leaf on some variable and creates two new children of the leaf, connecting each of them with an edge associated with one possible value of this variable (0 or 1). The second operator is a *merge*, which merges two leaves into a single one and introduces a new equality constraint on the parameter set.

Figure 6.2 shows the operators that can be used to construct the decision graph shown in Fig. 6.1(b) from an empty decision graph that contains only the marginal probability of the considered variable without any condition. A sequence of three splits and one merge leads to the resulting graph shown earlier in Fig. 6.1(b).

For the variables that can obtain more than two values, two versions of the split operator can be considered: (1) a complete split, which creates one child for each possible value of the variable (as above), and (2) a binary split, which creates one child corresponding to one particular value of the variable and another child for all the remaining values. These two operators are equivalent for binary variables. Of course, other alternatives can also be considered.

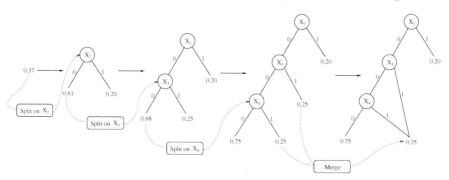

Fig. 6.2. Merge and split operators for decision graphs. The decision graph shown earlier in the text is constructed from an empty decision graph consisting of one leaf by a sequence of splits and merges

6.1.10 Constructing Bayesian Networks with Decision Graphs

The greedy algorithm for constructing Bayesian networks with decision graphs described in this section differs from the greedy algorithm for constructing traditional Bayesian networks presented in Sect. 3.3.2. The difference is that the algorithm for constructing the networks with decision graphs does not directly manipulate the network but it modifies the decision graphs corresponding to the variables in the network. The network B is initialized to an empty network that contains no edges. The decision graph G_i for each variable X_i is initialized to a single-leaf graph, containing only probabilities $p(X_i)$ as shown earlier in Fig. 6.2.

In each iteration, all operators (e.g., all possible merges and splits) that can be performed on all decision graphs G_i are examined. The operator that improves the score the most is performed on the corresponding decision graph. The operators that can be performed include (1) splitting a leaf of some decision graph on a variable that was not encountered on the path from the root to the leaf and (2) merging two leaves into a single leaf.

When performing a split, no cycles must be introduced to the network B. To guarantee that the final network remains acyclic, the network B can be updated after each split. After splitting a leaf of the graph G_i on X_j, an edge $X_j \to X_i$ is added to the network B if the edge is not already there. If a cycle would be introduced by the new edge, the split must not be allowed. The above restriction could be alleviated. The use of decision trees allows Bayesian multinets [49] with one or more distinguished variables. However, for the sake of computational efficiency and the simplicity of the implementation, we do not take this under consideration.

Modifying the decision graphs as opposed to modifying the network directly allows the greedy algorithm to make finer steps. Adding an edge into a Bayesian network and using a full conditional probability table to store the required probabilities corresponds to splitting all leaves of the decision graph

```
A greedy algorithm for network construction using decision graphs
  for each variable i {
     G(i) := single-node decision graph;
     B    := empty Bayesian network;
  }
  done := 0;
  repeat
     O := {};
     for each variable i {
       for each leaf l of G(i)
          add all applicable splits of l into O;
       for each pair of leaves l1 and l2 of G(i)
          add merge(l1,l2) into O;
     }
     if there exists an operation in O that improves score(B,G) {
        op = operation from O that improves score(B) the most;
        apply op to G(i);
        apply op to B;
     }
     else
        done := 1;
  until (done=1);
  return B, G;
```

Fig. 6.3. The pseudocode of the greedy algorithm for constructing a Bayesian network with decision graphs

corresponding to the terminal node of the edge on the variable corresponding to the initial node of the edge. However, by modifying only the decision graph, operators proceed in smaller and more specialized steps, which can lead to improving quality of the resulting model.

The following section focuses on the last key to hierarchy success – the preservation of alternative solutions.

6.2 Preservation of Alternative Candidate Solutions

This section focuses on methods for maintaining useful diversity in genetic and evolutionary algorithms. The preservation of alternatives is provided by encouraging local competition for the space in the population in some way; similar solutions compete with each other, whereas dissimilar solutions compete only rarely or never. Using an analogy from ecological systems, solutions that are similar share the same niche, and methods for preserving diversity are thus often referred to as *niching methods*.

The purpose of this section is to propose a niching method that will ensure the third key to hierarchy success – the preservation of alternative solutions – in hBOA. The section starts with an overview of niching methods. Methods that have been successfully used in traditional GAs are discussed in the context of BOA. One of these methods, restricted tournament replacement, is then adopted to ensure the preservation of alternative candidate solutions in hBOA.

6.2.1 Background of Niching

Niching methods localize competition in genetic and evolutionary algorithms so that candidate solutions compete in multiple niches and population diversity can be preserved for long periods of time. The purpose of niching in genetic and evolutionary computation is twofold:

(1) **Generate multiple solutions.** In some real-world applications it is important to find multiple solutions and let an expert, a simulation, or an experiment choose one of the solutions found or use multiple high-quality solutions. This is usually the case when there is insufficient knowledge to provide an exact definition of a single-objective evaluation function, when the fitness function does not fully determine solution quality but only focuses on several aspects of solution quality, or when for the sake of efficiency instead of using a complete fitness function one uses only its approximation that is more computationally efficient.

(2) **Preserve alternative solutions.** The reason for preserving multiple alternative solutions is that in some difficult problems one cannot clearly determine which candidate solutions are really on the right track until the optimization proceeds for a number of generations. Without effective niching, the population is subject to genetic drift (random effects, see Sect. 4.2.3), which may destroy some alternatives before it becomes clear whether or not they are the ones we are looking for.

There are three basic classes of niching methods:

(1) crowding,
(2) fitness sharing, and
(3) spatial separation.

Crowding modifies the selection procedure to localize competition by taking into account both the fitness as well as the genotype (solution string) or phenotype (semantics of the solution string) of competing candidate solutions [19, 34, 73, 103, 106]. *Fitness sharing* modifies the fitness landscape before selection so that the number of solutions corresponding to each niche is proportional to the fitness of that niche [37, 59, 86]. *Spatial separation* isolates several subpopulations of candidate solutions instead of keeping the entire population in one location [25, 26, 29, 64, 66, 111]. Candidate solutions can migrate between different locations (islands or demes) at certain intervals and allow the population at each location develop in isolation. Since competition among candidate solutions in each island is separated from other islands, solutions in different islands can differ significantly and diversity can be preserved for a long time. However, spatial separation does not force niching, it only *allows* niching to take place. That is why spatial separation is sometimes omitted in the context of niching.

Some related work studies the preservation of diversity from a different point of view. The primary goal of these techniques is not to preserve multiple solutions or alternative search regions, but to avoid *premature convergence*. Mauldin [105] proposed one such approach, which injects randomly generated candidate solutions into the current population at certain intervals. On the other hand, Baker [6] proposed a method that controls selection to prevent premature convergence. However, premature convergence is not the primary target of using niching in our work (although the preservation of alternatives and premature convergence are related). Various techniques for niching were also proposed in the area of multiobjective optimization [36, 43, 87, 163], but these methods are not applicable to single-criterion optimization.

The section continues with a brief overview of existing techniques in the three classes of niching methods: (1) crowding, (2) fitness-sharing, and (3) spatial separation. Subsequently, the section describes restricted tournament replacement used in hBOA.

Niching by Crowding

All niching methods localize competition in some way. In niching by crowding, competition is localized by modifying selection and replacement strategies.

Preselection [19], deterministic crowding [103], probabilistic crowding [106], and the gene invariant genetic algorithm (GIGA) [28] are applicable to GAs with two-parent recombination. In preselection [19], one candidate solution is created by crossing over two parent solutions, and the new solution replaces the inferior parent. Preselection encourages competition among similar candidate solutions, because an offspring and its parents usually share many similarities. Deterministic crowding [103] pairs each offspring with a more similar parent and the solutions within each pair compete. The offspring replaces the parent if it is better; otherwise the offspring is discarded. Probabilistic crowding [106] is a probabilistic extension of Mahfoud's deterministic crowding. In probabilistic crowding, the winner of the parent-offspring tournament is chosen according to the probabilities proportional to the fitness of the competitors. The gene invariant genetic algorithm (GIGA) [28] starts by selecting two solutions with fitness proportionate selection. After crossing over the selected solutions, the offspring replace their parents. In this fashion, the proportion of 0s and 1s in each position is preserved (gene invariance).

In crowding [34], for each new candidate solution, a subset of the original population is first selected. The new solution then replaces the most similar candidate solution in the selected subset. Earlier in the run only little will change compared to random replacement. However, as the run continues, candidate solutions will create groups of similar solutions, which compete for space with other members of the same group. Crowding was later used by Goldberg [52] for pipeline design by learning classifier systems.

Harik [73] proposed restricted tournament selection (RTS) as an extension of De Jong's crowding. Restricted tournament selection selects two parents at random according to a uniform distribution on the population. Two candidate solutions are then generated by applying crossover to the two selected parents. For each new candidate solution, a subset of the original population is first selected as in crowding. However, instead of automatically replacing the closest solution from the selected subset, the two solutions compete and the one that has a higher fitness wins. In this fashion, the selection step is performed by elitist replacement with a flavor similar to that of crowding. No extra selection operator is required. RTS was shown to perform well on a number of multimodal problems, locating all optima even on functions that are highly multimodal and difficult to solve, such as additively separable folder traps [73].

A number of techniques that *restrict mating* in some way to promote niching were proposed. Hollstein [85] required similar candidate solutions to form families and restricted mating to candidate solutions that share a family as long as their fitness improved. When the trend changes and the quality of a family decreases, crossbreeding across the families is allowed. Booker [11] discussed the need for restricted mating to prevent the formation of lethals. Perry [142] introduced multiple contexts in which the fitness function varied according to externally specified partial solutions that defined species. Candidate solutions could migrate between different contexts. This technique is mainly interesting for its biological background.

In the context of discrete PMBGAs, Pelikan and Goldberg [125] divided the population of selected parents in each generation into a specified number of clusters. A mixture of Gaussians was used to separate candidate solutions in the selected population using k-means clustering. Each cluster was processed separately and the offspring of each cluster was given a fraction of the new population of the size proportional to their average fitness. Niching was also investigated in the context of multi-objective PMBGAs [90, 98, 174].

De Jong's crowding and Harik's restricted tournament selection are two most straightforward approaches to incorporating niching into BOA. Both methods can be incorporated without affecting the model building or its utilization in BOA. However, it becomes difficult or impossible to use other crowding methods, such as Mahfoud's deterministic crowding, Mengshoel's probabilistic crowding, and Culberson's GIGA.

Niching by Fitness Sharing

The basic idea of fitness sharing is to create a number of artificial niches and give each niche a number of copies proportional to the quality of solutions in the niche. Although the motivation for fitness sharing dates back to Holland [83] and the k-armed bandit problem [34, 83], the first practical fitness-sharing method was developed by Goldberg and Richardson [59]. In fitness sharing, promising solutions are selected with the probability proportional to their fitness. However, before selection, the fitness of each candidate solution is modified according to

$$f'(X) = \frac{f(X)}{\sum_Y sh(d(X,Y))} ,$$

where $\sum_Y$ runs over all candidate solutions Y in the current population; $d(X,Y)$ is the distance between solutions X and Y; and $sh(d)$ is the sharing function. The sharing function defines the degree of similarity of two solutions that are located at distance d from each other as follows:

$$sh(d) = \begin{cases} 1 - \frac{d}{\sigma_{share}} & \text{if } d < \sigma_{share} \\ 0 & \text{otherwise} \end{cases} ,$$

where σ_{share} is the sharing threshold that defines the maximum distance between two candidate solutions that share a niche. After updating the fitness as described above, the number of copies of solutions in each niche is proportional to the average fitness of the niche.

For a successful application of fitness sharing, it is necessary to determine an adequate value of the sharing threshold σ_{share}. Deb and Goldberg [37] calculated the value of σ_{share} from the desired number of niches by dividing the problem domain (set of all candidate solutions) into a specified number of equally sized hyperspheres.

One of the drawbacks of fitness sharing is that it experiences difficulty with maintaining optima that are close to each other or those that are distributed irregularly. On the other hand, fitness sharing is capable of preserving all the optima for long periods of time (see for example the study of Horn [86] who analyzed the stability of sharing with two niches). Furthermore, unlike the approaches based on crowding, the fitness proportionate selection is capable of maintaining the size of each niche so that the number of copies of solutions in each niche is proportional to the average fitness of that niche.

Fitness sharing directly modifies the fitness before selection takes place. Consequently, the frequencies of partial solutions after selection are "disturbed" by sharing. Since those frequencies are the only input of BOA regarding the nonlinearities in the problem, BOA's capability of building a good model might suffer. That is why fitness sharing does not seem to be an appropriate approach to niching in BOA.

Niching by Spatial Separation

There are two reasons why spatial separation should be desirable in genetic and evolutionary computation: (1) In nature the populations are actually divided into a number of subpopulations that (genetically) interact only rarely or do not interact at all. (2) Separating a number of subpopulations allows for an effective parallel implementation and is therefore interesting from the point of view of computational efficiency. This section reviews and discusses niching methods based on spatial separation.

Spatial separation localizes competition by introducing a sort of geographical location of each candidate solution. Unlike in fitness sharing, in spatial separation the location of each solution does not depend on its genotype (the representation of the solution) or phenotype (the semantics of the solution). Amount of information exchange between the groups of candidate solutions from different locations is controlled by a specific strategy, which might depend on the distance or the relationship between the locations.

Much work in spatial separation was inspired by the shifting balance theory [182] and the theory of punctuated equilibria [40]. One approach is to divide the population into a number of subpopulations. Each subpopulation evolves in its own island and candidate solutions migrate between the islands at a certain rate. In this way, genetic material (partial solutions) is exchanged within each subpopulation often while its flow to other subpopulations is reduced. This approach was studied by Grosso [66], inspired mainly by the theory of Wright, and by Cohoon et al. [25], whose work is primarily inspired by the theory of Eldredge and Gould. Another approach introduces some kind of distance metric in the population and forces local competition and mating. This approach was studied by Gorges-Schleuter [64], Collins and Jefferson [26], Davidor [29], Mühlenbein [111], and others.

The primary drawback of spatial separation is that there is no direct mechanism that *forces* niching; instead, spatial separation *allows* for niching, because different islands can converge to different regions of the search space. Therefore, while spatial separation is a great approach for using multiprocessor architectures to process large populations in parallel, it does not seem to be a good approach for ensuring that useful diversity is maintained.

There are several additional problems with using spatial separation in BOA. Spatial separation requires isolated processing of a number of subpopulations. Although dividing the population might be beneficial for parallelization, it negatively affects BOA performance on single-processor machines, because the overall number of candidate solutions must be much larger to ensure that each subpopulation represents a large enough sample of promising solutions to find a good model. Furthermore, there must be many islands to ensure that useful diversity is maintained and (due to the independence of spatial separation on candidate solutions) one can never be sure that this indeed takes place.

6.2.2 The Method of Choice:
Restricted Tournament Replacement

The overview of niching methods discussed a number of methods that have been designed to ensure that useful diversity is maintained. Each niching

```
Restricted tournament replacement
  O := offspring;
  P := original population;

  for each offspring X from O {

    Y := random individual from P;
    dXY := distance(X,Y);

    for i=2 to window_size {
      Y' := random individual from P;
      dXY' := distance(X,Y');
      if (dXY'<dXY) {
        Y := Y';
        dXY := dXY';
      }
    }

    if (fitness(X)>fitness(Y))
      replace Y in P with X (Y is discarded);
    else
      discard X;
  }
```

Fig. 6.4. The pseudocode of the restricted tournament replacement (RTR)

method described in the above overview can be advantageous in some situations. So what niching method should be used in hBOA?

As follows from the discussion of each method, restricted tournament selection of Harik [73] and crowding of De Jong [34] seem to be the most suitable candidates. hBOA uses restricted tournament selection, because this niching method introduces an additional source of selection pressure that has proven advantageous according to a number of empirical results. However, since restricted tournaments are used for replacement in a generational GA (as opposed to the steady-state GA) and the primary selection pressure comes from selection, we call the used method *restricted tournament replacement* (RTR). RTR is outlined in Fig. 6.4.

The following section describes hBOA and discusses how hBOA differs from the original BOA described in Chap. 3.

6.3 Hierarchical BOA

Above we discussed the important components that must be incorporated into BOA to extend its applicability to difficult hierarchical problems. The purpose of this section is to describe the hierarchical BOA (hBOA) [127] and discuss the ways in which hBOA differs from the original BOA.

Recall that in BOA, the population is updated for a number of iterations (generations), each consisting of four steps: (1) selection, (2) model building, (3) model sampling, and (4) replacement. Each generation of hBOA also consists of the above four steps. However, instead of using traditional Bayesian networks, hBOA uses Bayesian networks with decision graphs to ensure the first two keys to hierarchy success – a proper decomposition and chunking. Therefore, the steps (2) and (3) of the basic BOA must be modified to incorporate decision graphs into the learning and sampling of Bayesian networks.

The second modification consists of using RTR as a replacement strategy. For each new candidate solution generated by sampling the learned model in step (3), a random subset of candidate solutions is first selected from the original population. The size of the selected subsets is fixed to a constant w, called the *window size*. The new solution is then compared to each candidate solution in the selected subset and the fitness of the most similar candidate solution of the subset is compared to that of the new solution. If the new solution is better, it replaces the most similar solution of the subset; otherwise, the new solution is discarded.

Based on our experience, a good heuristic is to set the window size to the number of bits in the problem:

$$w = n . \tag{6.5}$$

The reason for setting $w = n$ is that the number of niches n_{niche} that can be maintained in a population of candidate solutions is equal to some fraction of the population size: $n_{niche} = O(N)$. RTR with the window size w can maintain the number of niches that is equal to some fraction of the window size: $n_{niche} = O(w)$. Since in BOA, the population is expected to grow approximately linearly with the size of the problem, $N = O(n)$, to maximize the number of niches that can be maintained by BOA without affecting the population sizing, $w = O(n)$. It is also beneficial to restrict the window size to be much smaller than the population size; for example, $w \leq N/20$.

The basic hBOA procedure is outlined in Fig. 6.5.

```
Hierarchical BOA (hBOA)
  t := 0;
  generate initial population P(0);
  while (not done) {
    select population of promising solutions S(t);
    build Bayesian network B(t) with local struct. for S(t);
    sample B(t) to generate offspring O(t);
    incorporate O(t) into P(t) using RTR yielding P(t+1);
    t := t+1;
  };
```

Fig. 6.5. The pseudocode of the hierarchical Bayesian optimization algorithm

6.4 Experiments

In the previous chapter we have promised that if one incorporates chunking and niching into BOA, the resulting method should be capable of scalable optimization of difficult hierarchical problems. We have also designed a class of challenging hierarchical problems – hierarchical traps – that can be used to test whether an algorithm can solve hierarchical problems in a scalable manner.

The purpose of this section is to present experimental results of applying hBOA to the two types of hierarchical traps presented in Chap. 5. Additionally, the algorithm is tested on the HIFF function [181]. In all cases, hBOA passes the test and proves to be a scalable hierarchical optimizer. To show that hBOA is capable of discovering multiple alternative solutions, hBOA is also tested on a folded trap function with 32 global optima.

The section starts by introducing experimental methodology. Empirical results are then presented and discussed.

6.4.1 Methodology

For each problem instance, 30 independent runs are performed and hBOA is required to find the optimum in all the 30 runs. The performance of hBOA is measured by the average number of evaluations until the optimum is found. The population size for each problem instance is determined using the bisection method (see Fig. 3.8 on page 44) to be within 10% of the minimum population size for the algorithm to find the optimum in 30 independent runs.

Binary tournament selection with replacement is used in all experiments and the window size for RTR is set to the number of bits in a problem. Bayesian networks with decision graphs are used to model and sample candidate solutions and K2 metric with the term penalizing complex models is used to measure the quality of each candidate model.

6.4.2 Results

Figure 6.6 shows the performance of hBOA on the two hierarchical traps defined in Sect. 5.3.5. The number of bits in the first hierarchical trap is $n = 27$ to $n = 729$. For the largest problem ($n = 729$) only 7 runs are performed due to the increased computational requirements. The number of bits in the second hierarchical trap ranges from $n = 27$ to $n = 243$. In both cases, the number of evaluations is approximated by a function of the form $an^k \log n$, where a and k are set to fit the empirical results best. In both cases, the overall number of fitness evaluations grows subquadratically with problem size.

As shown in Fig. 6.7, similar results can be observed on the HIFF function [181] of $n = 16$ to $n = 512$ bits. The number of fitness evaluations is also approximated by a function of the form $an^k \log n$ and the overall number of fitness evaluations grows subquadratically with problem size.

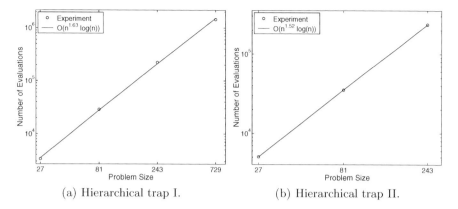

(a) Hierarchical trap I.   (b) Hierarchical trap II.

Fig. 6.6. The number of fitness evaluations until hBOA has found the optimum on the two hierarchical traps. The problem sizes range from $n = 27$ bits to $n = 729$ bits for the first trap, and $n = 27$ to $n = 243$ for the second trap. In both cases, the overall number of fitness evaluations grows subquadratically with the size of the problem

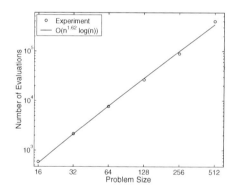

Fig. 6.7. The number of fitness evaluations until hBOA has found the optimum on the hierarchical if-and-only-if (HIFF) of $n = 16$ bits to $n = 512$ bits. The overall number of fitness evaluations grows subquadratically with the size of the problem

An additional experiment was performed on an additively separable folded trap function of order 6 to verify the effects of niching on the preservation of alternative solutions. More specifically, a 30-bit folded trap consisting of 5 copies of a 6-bit folded trap was used. The tested function provides an interesting test; there are 32 global optima and $3,200,000$ local optima. With the population of size $N = 1,500$, the algorithm can only afford to maintain all the global optima. Figure 6.8 shows that hBOA was capable of discovering all the global optima after about 43 generations. The global optima were further propagated and maintained until the end of the run of 110 generations. An interesting observation is that although all the optima are multiply represented

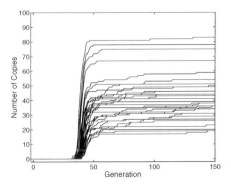

Fig. 6.8. The number of copies of different global optima of the bipolar function. There are 32 optima in this function and all 32 are multiply represented at the end of the run

at the end of the run, the number of copies of the optima differs significantly. The reason for this behavior is that unlike fitness sharing, restricted tournament replacement does not require that the size of each niche be proportional to the average fitness in that niche.

6.5 Scalability of hBOA on Hierarchical Problems

The results presented in the above section indicated that hBOA is capable of solving difficult hierarchical problems in a subquadratic number of evaluations. How does this relate to the BOA scalability theory presented in Chap. 4?

The convergence of hBOA on the hierarchical problems proceeds sequentially from the bottom to the top level. On each level, the correct building blocks on that level must be discovered and their competitors must be eliminated. The number of evaluations required by hBOA to discover the correct building blocks on each level can be upper-bounded by the overall number of fitness evaluations required to solve the problem on the current level only. Using the BOA scalability theory, on each level l the number of evaluations can be upper bounded by $O(n_l^{1.55})$, where n_l is the number of subproblems from the lower level, which serve as the basic building blocks (or bits) on the current level. For example, in the hierarchical trap, $n_l = n/3^{l-1}$. The number of levels is proportional to the logarithm of the problem size for all tested hierarchical problems. In particular, hierarchical traps contain $\log_3 n$ levels, and HIFF contains $\log_2 n$ levels. For hierarchical traps, the overall number of evaluations can be therefore bounded by

$$\sum_{l=1}^{\log_3 n} O\left(\left[\frac{n}{3^{l-1}}\right]^{1.55}\right) = O(n^{1.55} \log n) . \tag{6.6}$$

For HIFF, the same bound $O(n^{1.55}\log n)$ can be derived in a straightforward manner. Therefore, in both cases, the overall time to convergence is approximately equal to the number of levels times the time spent on a single-level problem of the same size. This result is confirmed by the empirical results.

If the number of levels grows faster than a logarithm of the problem size, the resulting bound is

$$E_{hboa} = O(n^{1.55}L(n)) , \tag{6.7}$$

where $L(n)$ denotes the number of levels in an n-bit problem. The bound given by the last equation could, however, change in some cases. Here we assumed that only two optima for each partition of the problem are preserved on each level to form the two building blocks that are to be juxtaposed on the next level. If the number of optima is not constant, the overall number of evaluations can be expected to increase due to the increased population size and number of generations. Furthermore, if the size of the partitions on the lowest level grows with the size of the problem, the population size will have to increase according to the BOA population sizing.

6.6 How Would Other Methods Scale Up?

We performed a number of experiments with other search methods, including traditional GAs, one- and two-bit deterministic hill climbers, the mutation-based stochastic hill climber, and the simulated annealing [91]. None of the tested algorithms was capable of solving hierarchical traps except for toy instances of at most $n = 27$ bits. That is why it was impossible to determine whether the growth of computational complexity of local search was exponential; nonetheless, this section argues that exponential complexity can be expected using traditional GAs as well as methods based on local search operators.

The reason for the poor performance of GAs with traditional recombination and no mutation is that without using an appropriate decomposition, the problem on any level of hierarchical traps requires exponentially large populations [176]. Therefore, even without hierarchy, GAs with traditional crossover yield intractable search. The following argument suggests that introducing mutation in GAs will not change the situation in this case (although mutation did yield polynomial search on additively separable traps, see Sect. 4.6).

Deterministic hill climbers start with a random string. In each iteration, they flip one or two bits of the current solution that result in the biggest increase in the fitness. If no more improvement is possible, the algorithms start over with a random starting point. Performing one- and two-bit flips is sufficient to discover some of the blocks 111 and 000 on the first level, but it fails to discover partial solutions on higher levels because the probability of hitting a partial solution decreases exponentially with its size. The only starting points from where the deterministic hill climber with one-bit flips is capable of reaching the optimum are the strings that contain at most one 0 in each partition of 3 bits in the first-level decomposition. The probability of generating such a starting point is given by

$$\left(\frac{1}{2}\right)^{\frac{n}{3}},$$

because there are 4 allowable instances of the 3 bits out of 8 instances total, and there are $n/3$ such blocks in an n-bit string. With two-bit flips, any instance of the 3 bits that contains at most two 0s is allowed, so the probability of generating a good starting point grows as

$$\left(\frac{7}{8}\right)^{\frac{n}{3}}.$$

Therefore, in both cases, exponential number of restarts must be performed to reach the optimum. Consequently, deterministic hill climbers are capable of solving only toy instances of 9 bits and they fail to solve hierarchical traps of size $n = 27$ or more even with tens of millions restarts. Note that because of the hierarchical nature of hierarchical traps, if we upper-bound the number of flips per iteration by any constant, the search becomes intractable.

A stochastic hill climber with bit-flip mutation also requires exponentially many restarts or exponentially many steps. This argument can be supported by the Markov chain analysis of the stochastic hill climber on additively separable traps. It can be shown that to flip particular k bits at the same time, bit-flip mutation requires approximately $O(n^k)$ trials [112]. If there are several such groups, it is most difficult to flip the last group of k bits and therefore the same complexity can be expected [112]. From any block of 3 bits, either 000 or 111 can be reached by flipping at most one bit; therefore, blocks 000 and 111 on the first level can be found in $O(n)$, yielding a string with half of the blocks equal to 000 and half of the blocks equal to 111. However, to find blocks 000000000 and 111111111 on the next level, three bits must be flipped at once (because of the deception of traps) and $O(n^3)$ trials are necessary. Continuing up a number of levels, the top level requires exponentially many trials to find the optimum.

An alternative approach to show the exponential number of trials of the stochastic hill climber is to bound the number of trials by a polynomial $O(n^k)$, and look at the starting points that allow the optimum to be found in the specified number of trials. This case is similar to the one where one or two bits are flipped deterministically, because $O(n^k)$ trials allow steps of size at most k flips. That results in an exponential number of restarts or finding a solution only up to a particular level. The maximum level that can be solved can be expected to be proportional to $\log k$.

The application of a cooling schedule of the simulated annealing relaxes the conditions somewhat; the groups of 3 and more bits do not have to be flipped at once, because the acceptance of each new solution is not deterministic. Indeed, with a slow enough cooling schedule, the simulated annealing is capable of solving a 27-bit hierarchical trap. Nonetheless, the complexity of the simulated annealing changes only marginally compared to hill climbers because local operators are still highly inefficient at moving toward the optimum on higher levels. As a result, it can be expected that even the simulated annealing requires exponentially many trials. To summarize, although hBOA was capable of solving hierarchical traps of size $n = 729$, its competitors could solve a problem of only $n = 27$ bits (and even for this problem, most competitors failed).

7

Hierarchical BOA in the Real World

The last chapter designed hBOA, which was shown to provide scalable solution for hierarchical traps. Since hierarchical traps were designed to test hBOA on the boundary of its design envelope, it was argued that if hBOA can scalably solve hierarchical traps, it should also be applicable to other hierarchical and nearly decomposable problems. Because many real-world problems are hierarchical or nearly decomposable, hBOA should be a promising approach to solving challenging real-world problems.

This chapter applies hBOA to two classes of real-world problems to confirm that decomposition and hierarchical decomposition are useful concepts in approaching real-world problems. Two classes of problems are considered: (1) two- and three-dimensional Ising spin glasses with periodic boundary conditions, and (2) maximum satisfiability (MAXSAT). The chapter shows that hBOA can achieve competitive or better performance than problem-specific approaches without the need for much problem-specific knowledge in advance. Additionally, the chapter relates the actual empirical performance of hBOA to the theory presented earlier in this book.

The chapter starts by defining the problem of finding ground states of Ising spin glasses. A hybrid algorithm created by combining hBOA with a simple deterministic local searcher is then applied to an array of two- and three-dimensional spin glasses with $\pm J$ couplings. Section 7.2 defines MAXSAT and discusses its difficulty. To improve performance, hBOA is again combined with local search. hBOA with local search is then applied to several benchmark instances of MAXSAT and compared to other MAXSAT solvers.

7.1 Ising Spin Glasses

The task of finding ground states of Ising spin glasses is a well known problem in physics. In the context of GAs, spin glasses are usually studied because of their interesting properties, such as symmetry and a large number of plateaus and local optima [116, 119, 132, 139, 179].

Martin Pelikan: *Hierarchical Bayesian Optimization Algorithm*, StudFuzz **170**, 131–146 (2005)
www.springerlink.com © Springer-Verlag Berlin Heidelberg 2005

The physical state of an Ising spin glass is defined by (1) a set of spins $(\sigma_1, \sigma_2, \ldots, \sigma_n)$, where each spin σ_i can obtain a value from $\{+1, -1\}$, and (2) a set of coupling constants J_{ij} relating pairs of spins σ_i and σ_j. A Hamiltonian specifies the energy of the spin glass as

$$H(\sigma) = -\sum_{\langle i,j \rangle} \sigma_i J_{ij} \sigma_j \ ,$$

where the sum runs over all connected spins denoted by $\langle i, j \rangle$. The task is to find configurations of spins called *ground states* for given coupling constants J_{ij} that *minimize* the energy of the spin glass. There are at least two ground states of each spin glass (the energy of a spin glass does not change if one inverts all the spins). In practice the number of ground states is usually much greater, and it often grows exponentially with the number of spins (problem size).

The problem of finding *any ground state* of a spin glass can be mapped to a well known combinatorial problem called *minimum-weight cut* (MIN-CUT) [3, 122]. However, since MIN-CUT is NP-complete [110], the mapping is not much helpful.

Here we consider a special case, where the spins are arranged on a two- or three-dimensional grid and each spin interacts with only its nearest neighbors in the grid (4 neighbors in 2D, 6 neighbors in 3D). Periodic boundary conditions are used to approximate the behavior of a large-scale system. Therefore, spins are arranged on a 2D or 3D toroid. Figure 7.1 shows an example two-dimensional spin glass with 9 spins arranged in a 3×3 toroid. Additionally, here we consider only Ising spin glasses with $\pm J$ couplings, where coupling constants are constrained to contain only two values, $J_{ij} \in \{+1, -1\}$.

The complexity of finding ground states of Ising spin glasses grows with the dimensionality of the problem. The 1D case is trivial and can be easily solved by starting in an arbitrary spin, and continuing spin by spin along the chain of coupled spins in one direction, always setting the spin value to minimize the energy of the system with respect to the coupling with the spin that was assigned last. In the worst case, one constraint will remain unsatisfied (one coupling will have positive contribution to the energy Hamiltonian). That is why we do not study 1D spin glasses in this paper. In the 2D case, several algorithms exist that can solve the restricted class of spin glasses in polynomial time [35, 47, 48, 89]. We will compare the best known algorithms to hBOA later in this section. In the 3D case, finding ground states of Ising spin glasses is NP-complete even if the coupling constants are restricted to $\{-1, 0, +1\}$ [10].

7.1.1 Methodology

In hBOA each state of the system is represented by an n-bit binary string, where n is the total number of spins. Each bit in a solution string determines the state of the corresponding spin: 0 denotes the state -1, 1 denotes the

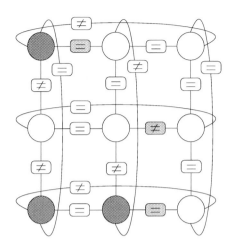

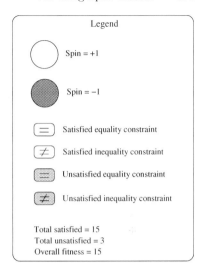

Fig. 7.1. An example two-dimensional Using spin glass of $n = 9$ spins, arranged on a toroid of size 3×3. Spins are shown as circles, non-zero coupling constants are shown as edges between pairs of spins. If $J_{ij} = -1$, the coupling constant introduces an equality constraint; if $J_{ij} = 1$, the coupling constant represents an inequality constraint. The more constraints are satisfied, the lower the energy of the underlying spin glass. The task is to maximize the number of satisfied constraints, which is equivalent to minimizing the energy of the underlying spin glass. The figure shows the values of the spins in one of the ground states (states with a minimum energy)

state $+1$. To estimate the scalability of hBOA on 2D Ising spin glasses, we tested hBOA on random instances for problems ranging from $n = 6 \times 6 = 36$ to $n = 20 \times 20 = 400$ spins, 1000 random instances for each problem size. To ensure that a correct ground state was found for each system, we verified the results for systems of size 6×6 to 16×16 with the ground states obtained by S. Sabhapandit and S. N. Coppersmith from the University of Wisconsin. For systems of size 18×18 and 20×20, the results we verified using the Spin Glass Ground State Server provided by the group of Prof. Michael Jünger.[1]

For each problem instance, 10 independent runs are performed and hBOA is required to find the optimum in all 10 runs. The performance of hBOA is measured by the average number of evaluations until the optimum is found. The population size for each problem instance is computed using the bisection method (see Fig. 3.8 on page 44 for the pseudocode) to be within 10% of the minimum population size that ensures convergence in all 10 runs. A parameter-less population sizing scheme [72] could be used to eliminate the need for specifying the population size in advance, which is expected to increase the total number of evaluations by at most a logarithmic factor [137, 138]. Binary tournament selection with replacement is used in all

---

[1] http://www.informatik.uni-koeln.de/ls_juenger/projects/sgs.html.

experiments and the window size for RTR is set to the number of bits (spins) in a problem, but it is bounded to be equal to at most 5% of the overall population size. Limiting the size of RTR significantly improved the performance of hBOA compared to the previously published results [124]. Bayesian networks with decision graphs are used and K2 metric with the term penalizing complex models is used to measure the quality of each candidate model as described in Pelikan et al. [134].

The performance of hBOA is improved by enhancing the initial population of candidate solutions using a simple heuristic based on a random walk through the grid. In each solution of the initial population, a random spin is first selected as a starting point for the random walk. Then, a random unvisited neighbor of the current spin is visited, setting its value to increase the fitness of the solution the most. The walk continues until all spins have been visited. If all neighbors of the current spin have already been visited, the walk continues in a randomly selected unvisited spin.

The performance of hBOA is also improved by combining hBOA with a local searcher referred to as the *discrete hill climber* (DHC). DHC is applied prior to the evaluation of each solution by flipping a bit that improves the solution the most; this is repeated until no more improvement is possible. For most constraint satisfaction problems including Ising spin glasses where the number of constraints for each variable is bounded by a constant, DHC increases the computational cost of each evaluation by at most $n \log n$ time steps. In practice, the increase in computational complexity is still significantly lower because only a few bits are flipped on average. Here, for instance, DHC does not increase the asymptotic computational complexity at all, while it decreases the required number of evaluations approximately tenfold [128].

Using local search often improves the performance of most PMBGAs, because the search can focus on local optima, which reveal more information about the problem than randomly generated solutions do. Furthermore, selectorecombinative search can focus its exploration on basins of attraction (peaks around each local optimum) as opposed to individual solutions. On the other hand, in some cases local search may cause premature convergence; nonetheless, we believe that this is rarely going to be the case with advanced algorithms such as BOA and hBOA.

Although the added heuristics improve the efficiency of hBOA, the asymptotic complexity of hBOA does not change much with the addition of the heuristics [124, 128]. Since the conclusions made from the experimental results are based on the asymptotic complexity, they are not a consequence of the additional heuristics.

7.1.2 Results

Figure 7.2a shows the number of evaluations for the hybrid combining hBOA with DHC (hBOA+DHC) on 2D spin glasses from $n = 6 \times 6$ to $n = 20 \times 20$

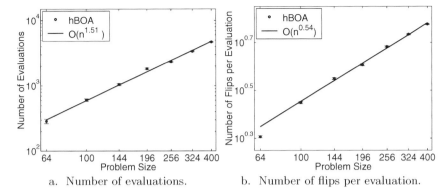

a. Number of evaluations.   b. Number of flips per evaluation.

Fig. 7.2. The number of evaluations and DHC flips per evaluation for hBOA+DHC on 2D Ising spin glasses (1000 random instances for each problem size). For each instance, 10 independent runs are performed and hBOA is required to find the optimum in all 10 runs

and Fig. 7.2b shows the number of flips per evaluation for the same set of instances. The number of evaluations appears to grow polynomially as $O(n^{1.51})$, whereas the number of flips per evaluation appears to grow as $O(n^{0.54})$. The performance of hBOA is thus even slightly better than predicted by hBOA scalability theory, which estimates the number of evaluations for difficult hierarchical problems as $O(n^{1.55} \log n)$. Moreover, the average number of flips per evaluation indicates that the asymptotic complexity of one evaluation will remain approximately linear with the problem size and it will thus not increase because of DHC.

7.1.3 Comparison with Other Black-Box Optimizers

Figure 7.3 compares performance of hBOA with that of simple GAs with one-point and uniform crossover (Fig. 7.3a), and the univariate marginal distribution algorithm (UMDA) (Fig. 7.3b), which uses a probability vector to sample new candidate solutions (a Bayesian network with no edges). The experimental methodology was identical to that for hBOA including DHC as a local searcher and a random walk for initializing the population. The figure shows that the number of evaluations required by GAs with uniform crossover and UMDA grows exponentially with the problem size. The number of evaluations for GAs with one-point crossover also starts to grow exponentially fast; however, the exponent is slowly decreasing and it can be expected that the number of evaluations can be bounded by a polynomial of order at least 3.56. An intuitive explanation of this behavior is that in these experiments, the grid is coded with binary strings so that one-point crossover does not break important interactions often. Such an encoding is the best encoding one could expect for one-point crossover; any perturbation of this encoding would result

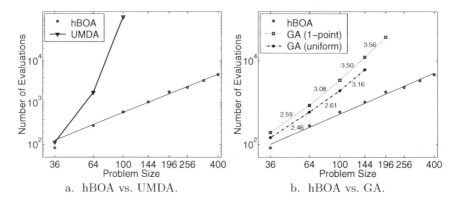

a. hBOA vs. UMDA.   b. hBOA vs. GA.

Fig. 7.3. hBOA+DHC vs. UMDA+DHC and the simple GA+DHC on 2D Ising spin glasses (1000 random instances for each problem size). For both GA variants, slopes are displayed adjacent to the corresponding line segments to highlight exponential complexity

in the behavior similar to that of the GA with uniform crossover. But no matter whether the performance in this ideal case is polynomial or not, the performance in all cases seems to be either exponential or qualitatively worse than that of hBOA.

We have also tried a (1+1)-ES with bit-flip mutation, which was not able to solve any but smallest problems of $n = 6 \times 6 = 36$ and $n = 8 \times 8$ bits. Already for problems of size $n = 10 \times 10 = 100$ bits, (1+1)-ES was not able to find the optimum in most cases even after hours of computation, which suggests exponential performance as well. Exponential performance was also the case for DHC with random restarts if trapped in a local optimum. Similar performance can be expected for other methods based on local search. Exponential scaleup was shown also for state-of-the-art Monte Carlo methods [32].

7.1.4 Comparison with Problem-Specific Methods

To compare the computational complexity of hBOA with problem-specific methods for solving spin glasses, let us first compute the overall computational complexity of hBOA+DHC on the tested spin glass instances. The number of evaluations appears to grow as $O(n^{1.51})$, where approximately $O(n^{0.91})$ comes from the population size, and $O(n^{0.6})$ comes from the number of generations. Every evaluation can be done in $O(n)$ steps (including DHC, which in this case takes approximately $O(\sqrt{n} \log n)$ steps per evaluation). The overall time spent in evaluation is thus bounded by $O(n^{2.51})$. The complexity of model building in each generation can be bounded by $O(kn^2N)$, where N is the population size and k is the depth of decision trees. Assuming that the depth of decision trees grows at most logarithmically with the size of the problem, we get $O(n^{3.51} \log n)$ steps for model building overall. The assumption on the logarithmic bound on the decision-tree depth can be expected to hold due to the linear upper bound on the population size. Replacement using RTR can also be bounded by $O(n^{3.51})$ steps, whereas selection can be done in only $O(n^{2.51})$ steps total. The primary source of computational complexity is thus the model building and RTR, which can be bounded by $O(n^{3.51} \log n)$ steps overall. This result is better than the results reported by Pelikan [124]; this qualitative improvement was due to the limited size of RTR window size, which is now limited to be at most 5% of the actual population size.

There are several problem-specific algorithms that attempt to solve the above special case of 2D spin glasses (e.g., Kardar and Saul [89], De Simone et al. [35], Galluccio and Loebl [47], Galluccio and Loebl [48]). Most recently, Galluccio and Loebl [47, 48] proposed an algorithm for solving spin glasses in $O(n^{3.5})$ for all graphs with bounded genus (two-dimensional toroids are a special case of graphs with bounded genus). So, the overall time complexity of the best currently known algorithm for the considered class of spin glasses is $O(n^{3.5})$. However, it is important to note that the method of Galluccio and Loebl can find the distribution of states over the entire energy spectrum, while hBOA focuses on only finding one or several configurations with minimum energy. That is why the method of Galluccio and Loebl solves in fact a more difficult problem.

The above results indicate that hBOA performs similarly as the best problem-specific approach; the time complexity of hBOA can be estimated as $O(n^{3.51} \log n)$, whereas the time complexity of the method of Galluccio and Loebl is $O(n^{3.5})$. Despite that hBOA does not use any problem-specific knowledge except for the evaluation of suggested states of the system and the method of Galluccio and Loebl fully relies on the knowledge of the problem structure and its properties, hBOA is capable of providing almost the same asymptotic complexity. But hBOA is not competitive only with respect to the asymptotic complexity; comparing the running times will reveal that hBOA is capable of outperforming the method of Galluccio and Loebl even with respect to the overall running time [124], so there seem to be no large constants hidden in $O(\cdot)$ notation for the computational complexity of hBOA. As previous results suggest, the asymptotic complexity can be expected to be retained even without the use of the random-walk heuristic used for population initialization and local search, but the factor by which the overall time complexity increases does not allow for such a detailed complexity analysis.

7.1.5 From 2D to 3D

Finding a ground state of 3D spin glasses is NP-complete even for coupling constants restricted to $\{-1, 0, +1\}$ [10]. Despite that, some positive results in solving 3D spin glasses are reported for instance in Hartmann [78]. Here we focus on a special case where the coupling constants are restricted to be $\{-1, +1\}$. Since hBOA does not explicitly use the dimensionality of the underlying spin-glass problem, it is straightforward to apply hBOA+DHC

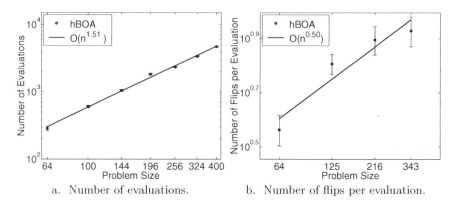

a. Number of evaluations.   b. Number of flips per evaluation.

Fig. 7.4. The number of evaluations and DHC flips per evaluation for hBOA+DHC on 3D Ising spin glasses (8 random instances for each problem size)

to 3D spin glasses. In this section we present preliminary results on several instances of 3D Ising spin glasses.

To test the scalability of hBOA+DHC, eight random spin glasses on a 3D cube with periodic boundary conditions have been generated for systems of size from $n = 4 \times 4 \times 4 = 64$ to $n = 7 \times 7 \times 7 = 343$ spins. To verify whether the found state actually represents the ground state, hBOA+DHC was first run on each instance with an extremely large population of orders of magnitude larger than the expected one. After a number of generations, the best solution found was assumed to represent the ground state. Similar results were obtained for other 3D spin glass instances provided by Alexander Hartmann from the University of Goettingen.

Figure 7.4 shows the number of evaluations (Fig. 7.4a) and the number of flips per evaluation (Fig. 7.4b) until hBOA+DHC found the ground state of the tested 3D Ising spin-glass instances. The overall number of evaluations appears to grow slightly faster than $O(n^{2.91})$. Although we do not allow coupling constants to be 0 and thus it is not clear whether the tested class of spin glasses is NP complete, the number of evaluations can be expected to grow faster than polynomially. However, these results hold a big promise and we are currently investigating how to use these results as the starting point for solving enormously large 3D instances that are intractable with any other method. One of the ways for further improvement of hBOA performance is to couple hBOA with a more sophisticated local searcher for spin glasses, such as the cluster exact approximation (CEA) [77]. To solve such enormously large instances, hBOA will have to be run on massively parallel computers to distribute computation among multiple processors. Preliminary results indicate that for complex spin glass instances almost linear speedups can be achieved even with tens or hundreds of computers and a distributed memory architecture [120].

7.2 Maximum Satisfiability (MAXSAT)

The task of finding an interpretation of propositions that maximizes the number of satisfied clauses of a given propositional logic formula expressed in conjunctive normal form, or MAXSAT, is an important problem of complexity theory and artificial intelligence. Since MAXSAT is NP-complete in its general form, there is no known algorithm that can solve MAXSAT in worst-case polynomial time.

In the context of GAs, MAXSAT is usually used as an example class of problems that *cannot* be efficiently solved using GAs based on selection and recombination [145], although some positive results were reported with adaptive fitness [65]. One of the reasons for poor GA performance appears to be that short-order partial solutions lead away from the optimum (sometimes in as many as 30% of propositions) as hypothesized by Rana and Whitley [145]. Another reason is that the interaction structure in most MAXSAT instances is complex and traditional recombination operators can thus be expected to yield inferior performance. Since hBOA outperforms GAs on complex decomposable and hierarchically decomposable problems, it would be interesting to apply hBOA to MAXSAT in order to find out whether hBOA is able to make difference and show that selection and recombination can be successfully used to solve MAXSAT as well.

Here we consider logic formulas in conjunctive normal form with clauses of length at most k; formulas in this form are called k-CNF formulas. A CNF formula is a *logical and* of clauses, where each clause is a *logical or* of k or less literals. Each literal is either a proposition or a negation of a proposition. An example 3-CNF formula defined with 5 propositions X_1, X_2, X_3, X_4, and X_5 is

$$(X_5 \vee X_1 \vee \neg X_3) \wedge (X_2 \vee X_1) \wedge (\neg X_4 \vee X_1 \vee X_5) .$$

An interpretation of propositions assigns each proposition either true or false; for example, $(X_1 = \text{true}, X_2 = \text{true}, X_3 = \text{false}, X_4 = \text{false}, X_4 = \text{true})$ is an interpretation of X_1 to X_5. The task is to find an interpretation that maximizes the number of satisfied clauses in the given formula. For example, the assignment $(X_1 = \text{true}, X_2 = \text{true}, X_3 = \text{true}, X_4 = \text{true}, X_5 = \text{true})$ satisfies all the clauses in the above formula, and is therefore one of the optima of the corresponding MAXSAT problem. MAXSAT is NP complete for k-CNF if $k \geq 2$. However, it is possible to check whether a 2-CNF formula is satisfiable (all clauses can be satisfied) in polynomial time.

7.2.1 Methodology

In hBOA, each candidate solution represents an interpretation of propositions in the problem. Each bit in a solution string corresponds to one proposition; true is represented by 1, false is represented by 0. The fitness of a solution is equal to the number of satisfied clauses given the interpretation encoded by the solution. The deterministic hill climber with one-bit flips is used to reduce the population-sizing requirements and improve the efficiency of the search. The hill climber flips the bit that improves the current solution the most until no more improvement is possible. DHC for MAXSAT is often called GSAT in the machine learning community [169]. Each iteration of GSAT can be performed in one pass through the formula.

For each problem instance, 30 independent runs are performed and hBOA is required to find the optimum in all the 30 runs. The performance is measured by the average number of evaluations until the optimum is found. The population size is determined empirically using the bisection method (see Fig. 3.8 on page 44 for the pseudocode) to be within 10% of the minimum population size required for convergence to the optimum in all the runs. Binary tournament selection with replacement is used in all experiments and the window size for RTR is set to the number of bits (propositions) in the problem. Bayesian networks with decision graphs are used and K2 metric with the term penalizing complex models is used to measure model quality.

7.2.2 Other MAXSAT Solvers Included in Comparison

Two MAXSAT solvers are included in the comparison: GSAT and WalkSAT. In addition to GSAT and WalkSAT, the comparison includes a powerful SAT solver called Satz [99]. Since Satz is a SAT solver, it cannot actually solve MAXSAT but it can only verify whether a given formula is satisfiable or not.

GSAT [169] is a deterministic hill climber based on one-bit flips. GSAT generates the initial interpretation at random according to uniform distribution over all interpretations. In each iteration, GSAT changes the interpretation of the proposition that leads to a largest increase in the number of satisfied clauses. If no more improvement of the current solution is possible, GSAT is restarted with a random solution.

WalkSAT extends GSAT to incorporate random changes. In each iteration, WalkSAT performs the greedy step of GSAT with the probability p; otherwise, one of the propositions that are included in some unsatisfied clause is randomly selected and its interpretation is changed. Best results are usually obtained with $p = 0.5$, where both GSAT and the random perturbation are applied with the same probability. However, the optimal choice of p might change from application to application. We observed that for instances tested in this chapter, WalkSAT performs best when $p = 0.5$.

Satz [99] is an extension of the Davis-Putnam-Logemann-Loveland algorithm [31]. Satz uses resolution (a sound and complete proof procedure for CNF) and several simple heuristics to find a satisfying interpretation for the input formula or a proof that the formula is unsatisfiable. As was mentioned above, Satz does not actually solve MAXSAT; instead of finding an interpretation that maximizes the number of satisfied clauses in the input formula, Satz only verifies whether the formula is satisfiable or not. Verifying satisfiability of a formula is an easier task than that of finding an assignment that maximizes the number of satisfied clauses.

7.2.3 Tested Instances

Two types of MAXSAT instances are tested: (1) random satisfiable 3-CNF formulas, and (2) instances of combined-graph coloring translated into MAXSAT. All tested instances have been downloaded from the Satisfiability Library SATLIB[2].

Instances of the first type have been randomly generated satisfiable 3-CNF formulas. Only instances from the phase transition for MAXSAT and 3-CNF [21] are considered, where the number of clauses is equal to $4.3n$ (n is the number of propositions). *Random* problems in the phase transition are known to be the most difficult ones for most MAXSAT heuristics [21].

There are two approaches to ensuring that random formulas are satisfiable. The first approach generates a satisfying interpretation first, and then generates only such clauses that are satisfied in the specified interpretation; these instances are called *forced satisfiable instances*. The second approach is to generate formulas at random first (with uniform distribution), and then filter out unsatisfiable instances using some of the complete algorithms such as Satz; these instances are called *unforced filtered satisfiable instances*. It is known that forced satisfiable instances are easier than unforced filtered satisfiable instances. Here, hard instances are used, which are obtained by filtering (formulas are unforced filtered). Despite generating problem instances from the phase transition, all tested instances are rather easy for both WalkSAT and Satz.

Instances of the second type have been generated by translating graph-coloring instances to MAXSAT. In graph coloring, the task is to color the vertices of a given graph so that no connected vertices share the same color. The number of colors is bounded by a constant. Every graph-coloring instance can be mapped into a MAXSAT instance by introducing one proposition for each pair (color, vertex), and creating a formula that is satisfiable if and only if exactly one color is chosen for each vertex, and the colors of the vertices corresponding to each edge are different.

Here, graph-coloring instances translated into MAXSAT instances are generated by combining regular ring lattices and random graphs with a specified number of neighbors [51]. Combining two graphs consists of selecting (1) all edges that overlap in the two graphs, (2) a random fraction $(1 - p)$ of the remaining edges from the first graph, and (3) a random fraction p of the remaining edges from the second graph. By combining regular graphs with random ones, the amount of structure in the resulting graph can be controlled; as p decreases, graphs are becoming more regular (for $p = 0$, the resulting graph is a regular ring lattice).

---

[2] http://www.satlib.org/

Table 7.1. An overview of MAXSAT instances used in the experiments

SATLIB Archive	Vars.	Clauses	Description
uf20-91.tar.gz	20	91	Random 3-CNF instances
uf50-218.tar.gz	50	218	from the phase transition
uf75-325.tar.gz	75	325	region ($c = 4.3n$)
uf100-430.tar.gz	100	430	Most instances easy for
uf125-538.tar.gz	125	538	WalkSAT and Satz
uf150-645.tar.gz		150	645
sw100-8-1p1-c5.tar.gz	500	3100	Instances obtained by
sw100-8-1p2-c5.tar.gz	500	3100	translating graph
sw100-8-1p3-c5.tar.gz	500	3100	coloring to MAXSAT
sw100-8-1p4-c5.tar.gz	500	3100	Graphs created by
sw100-8-1p5-c5.tar.gz	500	3100	combining regular
sw100-8-1p6-c5.tar.gz	500	3100	lattices with random
sw100-8-1p7-c5.tar.gz	500	3100	graphs are considered
sw100-8-1p8-c5.tar.gz	500	3100	Graphs are 5-colorable
			Most instances difficult for
			WalkSAT, some difficult for
			Satz

For small values of p (from about 0.003 to 0.03), MAXSAT instances of the second type are extremely difficult for WalkSAT and other methods based on local search. On the other hand, for higher values of p, some instances are extremely difficult for Satz and other complete methods. All instances are created from graphs of 100 vertices and 400 edges that are colorable using 5 colors, and each coloring is encoded using 500 binary variables (propositions).

Table 7.1 summarizes tested MAXSAT instances and their basic properties.

7.2.4 Results on Random 3-CNF Satisfiable Instances

Figure 7.5 shows the number of evaluations for hBOA+GSAT on randomly generated (unforced filtered) 3-CNF MAXSAT instances. Ten random instances are tested for each problem size. The growth of the number of evaluations can be approximated by a polynomial $O(n^{3.45})$, although an exponential growth can be expected in the worst case.

How does the performance of hBOA+GSAT compare to that of other approaches? Figure 7.6(a) compares the performance of hBOA+GSAT with that of GSAT alone. GSAT is capable of solving only the simplest instances of up to $n = 75$ variables, because the computational time requirements of GSAT grow extremely fast. Already for instances of $n = 100$ variables, GSAT could not find an optimal interpretation even after days of computation. The increasing slope of GSAT complexity in logarithmic scale indicates that the number of evaluations required by GSAT grows exponentially fast with problem size.

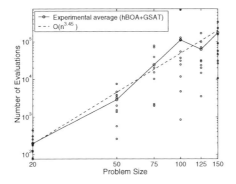

Fig. 7.5. Results of hBOA+GSAT on MAXSAT for randomly generated 3-CNF satisfiable formulas (unforced). The problem instances were downloaded from SATLIB

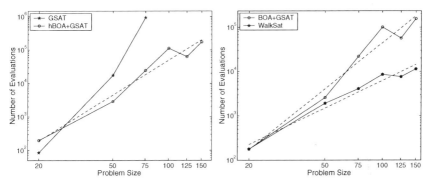

(a) Comparison of hBOA+GSAT with GSAT alone.   (b) Comparison of hBOA+GSAT with WalkSAT.

Fig. 7.6. Comparison of the performance of hBOA+GSAT with the performance of GSAT alone and WalkSAT on MAXSAT for randomly generated 3-CNF satisfiable formulas (unforced). The problem instances were downloaded from SATLIB. hBOA+GSAT outperforms GSAT alone; however, hBOA+GSAT is outperformed by WalkSAT

Therefore, GSAT alone cannot solve the problem efficiently, although it improves the efficiency of hBOA when used in the hybrid hBOA+GSAT.

Figure 7.6(b) compares the performance of hBOA+GSAT with that of WalkSAT. The figure indicates that the performance of WalkSAT is better than that of hBOA+GSAT both in the magnitude and in the growth. Therefore, a simple randomization of GSAT performs better than the sophisticated bias of hBOA+GSAT. Nonetheless, although hBOA uses selectorecombinative bias based on problem decomposition and hierarchical problem decomposition, it is capable of competing with local search on problem instances that are rather easy for local search.

Satz can verify satisfiability of all tested instances relatively fast. That indicates that the tested instances are hard only for GSAT, which appears to require exponential time to solve these instances. The performance of hBOA+GSAT is comparable to (but worse than) that of WalkSAT, and it is qualitatively better than that of GSAT.

7.2.5 Results on Combined-Graph Coloring

Randomly generated 3-CNF instances are rather easy for most tested algorithms. Nonetheless, real-world problems are not random, most real-world problems contain a considerable amount of regularity. Combined-graph coloring described in Sect. 7.2.3 provides an interesting class of problems, where regularity is combined with randomness. By controlling the relative amounts of structure and randomness, interesting classes of problems can be generated. This section tests the algorithms that performed relatively well on random 3-CNF, and applies these algorithms to combined-graph coloring (translated into MAXSAT).

Although regular ring lattices ($p = 0$) can be solved by WalkSAT efficiently [51], introducing even a slight perturbation to a regular ring lattice by combining it with a random graph severely affects WalkSAT performance. More specifically, WalkSAT is practically unable to solve any instances with $p \leq 2^{-5}$ even with a very large number of restarts and trials. For these problem instances, GSAT performance is also poor.

On the other hand, hBOA+GSAT is capable of solving all these instances despite their large size (500 variables). Table 7.2 shows the number of evaluations for hBOA+GSAT on several instances that are practically unsolvable by WalkSAT. WalkSAT has not been able to solve any of these instances even when allowed to check over 40 million interpretations (when the runs are terminated).

Satisfiability of most instances of the second type is easy to verify with Satz. However, there are several instances for which Satz performs poorly. We selected some of those instances, and applied hBOA+GSAT to them. Table 7.3 shows the performance of hBOA+GSAT and Satz on those instances. The performance of Satz is measured by the number of branchings, which corresponds to the number of decisions that Satz must make until it outputs the final answer. Since Satz is deterministic, only one run of Satz suffices for each instance. For Satz-hard instances shown in the table, hBOA+GSAT proves its robustness, because it is capable of solving all these instances in time comparable to that on other instances.

7.2.6 Discussion

There are several important observations regarding the performance of the hybrid hBOA+GSAT on the tested MAXSAT instances.

Table 7.2. The number of evaluations for hBOA+GSAT on MAXSAT instances that are practically unsolvable by WalkSAT. The instances are generated by translating the graph coloring of 5-colorable combined graphs. All instances have 500 variables and 3600 clauses, and they were downloaded from SATLIB

Instance	p	hBOA+GSAT Evaluations	WalkSAT Evaluations
SW100-8-5/sw100-1.cnf	2^{-5}	1,262,018	>40,000,000
SW100-8-5/sw100-2.cnf	2^{-5}	1,099,761	>40,000,000
SW100-8-5/sw100-3.cnf	2^{-5}	1,123,012	>40,000,000
SW100-8-6/sw100-1.cnf	2^{-6}	1,183,518	>40,000,000
SW100-8-6/sw100-2.cnf	2^{-6}	1,324,857	>40,000,000
SW100-8-6/sw100-3.cnf	2^{-6}	1,629,295	>40,000,000
SW100-8-7/sw100-1.cnf	2^{-7}	1,732,697	>40,000,000
SW100-8-7/sw100-2.cnf	2^{-7}	1,558,891	>40,000,000
SW100-8-7/sw100-6.cnf	2^{-7}	1,966,648	>40,000,000
SW100-8-7/sw100-7.cnf	2^{-7}	1,222,615	>40,000,000
SW100-8-8/sw100-1.cnf	2^{-8}	1,219,675	>40,000,000
SW100-8-8/sw100-2.cnf	2^{-8}	1,537,094	>40,000,000
SW100-8-8/sw100-6.cnf	2^{-8}	1,650,568	>40,000,000
SW100-8-8/sw100-7.cnf	2^{-8}	1,287,180	>40,000,000

Table 7.3. The number of evaluations for hBOA+GSAT on MAXSAT instances that are extremely hard for Satz. The instances are generated by translating the graph coloring of 5-colorable combined graphs. All instances have 500 variables and 3600 clauses, and they were downloaded from SATLIB

Instance	p	hBOA+GSAT Evaluations	Satz Branchings
SW100-8-1/sw100-1.cnf	2^{-1}	2,927,182	314,051,251
SW100-8-1/sw100-15.cnf	2^{-1}	3,578,068	12,350,045
SW100-8-2/sw100-44.cnf	2^{-2}	1,983,397	16,017,146

hBOA+GSAT has outperformed GSAT alone on all problem instances; not surprisingly, hBOA is capable of supplying much better starting points for GSAT than random restarts do. However, on the class of randomly generated 3-CNF, the hybrid hBOA+GSAT has been outperformed by a randomized GSAT called WalkSAT.

On the other hand, for those problem instances that are practically unsolvable by any local search methods included in the comparison (WalkSAT and GSAT), hBOA+GSAT retains efficient performance. In particular, MAXSAT instances obtained by translating graph coloring of graphs with a large amount of structure and a little amount of randomness cannot be solved by GSAT or WalkSAT even after tens of millions of evaluations, whereas hBOA+GSAT is capable of solving all these problems in fewer than two million evaluations.

Therefore, hBOA+GSAT can solve those instances that are easy for local search (random 3-CNF), but it is not limited to those instances – it can solve also problems that are practically unsolvable by local search alone.

Although Satz can verify satisfiability of most formulas efficiently, for several instances the performance of Satz is extremely poor. As shown in Table 7.3, hBOA+GSAT is capable of solving even those Satz-hard instances in time comparable to that on other instances. Nonetheless, recall that Satz does not actually solve MAXSAT. Satz can only find a satisfying interpretation of the formula or a proof that the formula is not satisfiable, but it cannot find the best interpretation for an unsatisfiable formula; hBOA, GSAT, and WalkSAT can.

To summarize, hBOA can solve an array of MAXSAT problems of different structure without requiring any information about the problem except for a measure of performance of each candidate solution and the number of bits in candidate solutions. On some MAXSAT problems, hBOA is outperformed by specialized heuristics; however, hBOA proves its robustness by solving all problems that it was applied to, whereas other methods are very good on some problems and very bad on other problems. Furthermore, note that the same algorithm was applied to onemax, composed traps, exponentially scaled deceptive problems, hierarchical traps, spin glasses, and MAXSAT. hBOA is therefore robust not only with respect to specialized classes of problems such as spin glasses or MAXSAT, but it provides scalable solution for a broad spectrum of qualitatively different problem classes, as long as these problems are decomposable or hierarchically decomposable.

8

Summary and Conclusions

The purpose of this chapter is to provide a summary of main contributions of this work and outline important conclusions.

8.1 What Has Been Done

A summary of the major results of this book follows:

Bayesian optimization algorithm (BOA). The book proposed the Bayesian optimization algorithm (BOA), which replaces traditional variation operators of GAs by (1) building a Bayesian network as a model of promising solutions and (2) generating new solutions by sampling the built network. Building a Bayesian network allows BOA to automatically discover regularities in the problem. Sampling the constructed network uses information encoded by the model to effectively explore the search space. BOA can reliably solve hard problems of bounded difficulty in a quadratic or subquadratic number of function evaluations with respect to the problem size.

Scalability theory of BOA. Theory was developed that estimates the number of fitness evaluations until convergence of BOA on decomposable problems of bounded difficulty. The number of fitness evaluations was computed by (1) approximating an adequate population size for reliable convergence to the optimum, (2) estimating the number of generations until convergence, and (3) making a product of these two quantities. The theory confirmed that the number of evaluations until convergence to the optimum on problems of bounded difficulty grows subquadratically or quadratically with the problem size, depending on the scaling of subproblems.

Hierarchy for complexity reduction. Hierarchical decomposition allows for a scalable solution of those problems that are not decomposable on

Martin Pelikan: *Hierarchical Bayesian Optimization Algorithm*, StudFuzz **170**, 147–149 (2005)
www.springerlink.com   © Springer-Verlag Berlin Heidelberg 2005

a single level but that can be solved by decomposition over multiple levels of difficulty. The book identified three important concepts that must be incorporated into an optimizer to scalably solve difficult hierarchical problems. These concepts comprise the three keys to hierarchy success: (1) proper decomposition, (2) chunking, and (3) preservation of alternative solutions. On each level, the problem must be decomposed properly so that the algorithm is not misled by salient nonlinearities on the current level. Best partial solutions of the subproblems on each level must be represented in a compact way and a mechanism for chunking partial solutions must be introduced so that these can be juxtaposed effectively on higher levels. Finally, alternative partial solutions to the subproblems on each level of decomposition must be preserved until it becomes possible to eliminate some of the alternatives.

Hierarchical traps. To develop competent optimization methods, it is important to design problems that can be used to effectively test developed techniques on the boundary of their design envelope. The design of difficult hierarchical problems can be guided by the three keys to hierarchy success, so that any algorithm that is incapable of tackling the three keys to hierarchy success will fail at solving the proposed class of problems efficiently. The book proposed hierarchical traps to provide the test of fire for optimization methods that attempt to exploit hierarchical decomposition. Hierarchical traps are practically unsolvable if the three keys to hierarchy success are not dealt with properly.

Hierarchical BOA. Hierarchical BOA (hBOA) extends BOA by (1) using local structures for a more compact representation of local probability distributions in Bayesian networks and (2) using restricted tournament replacement to incorporate new candidate solutions into the original population of candidate solutions. hBOA was shown to solve hierarchical traps and other hard problems in a subquadratic or quadratic number of evaluations with respect to the problem size.

Experiments on artificial and real-world problems. A number of experiments to verify the scalability of BOA and hBOA were performed. Three fundamentally different sets of problems were used: (1) boundedly difficult decomposable problems, (2) difficult hierarchical problems, and (3) real-world problems. Empirical results on boundedly difficult problems – onemax, composed traps, composed deceptive, and exponentially scaled deceptive functions – indicated that BOA can solve the class of problems decomposable into subproblems of bounded order in a scalable manner. Empirical results on hierarchically difficult problems – hierarchical traps and HIFF – indicated that hBOA is capable of solving difficult hierarchical problems in a scalable manner. Empirical results on two classes of real-world problems – 2D and 3D Ising spin-glass systems and MAXSAT – showed that not only can hBOA compete with state-of-the-art methods that are designed to solve these classes of problems, but that it can oftentimes outperform those methods.

8.2 Main Conclusions

A scalable black-box optimization algorithm capable of automatic discovery and effective exploitation of single-level and hierarchical problem decomposition exists and is ready for application. The Bayesian optimization algorithm (BOA) and the hierarchical Bayesian optimization algorithm (hBOA) can solve a broad class of problems as long as these problems are decomposable on a single or multiple levels of difficulty. For a successful application of BOA and hBOA, no prior knowledge about the problem is required except for (1) the number of decision variables and (2) a measure for evaluating candidate solutions.

There are only a few parameters that must be set in BOA and hBOA. In fact, the only parameter that really matters is the population size, because all the remaining parameters change the performance only marginally. The population size can also be eliminated by using the parameter-less population-sizing scheme of Harik and Lobo [72]. Promising results with the parameter-less hBOA were reported by Pelikan and Lin [137]. Therefore, the developed algorithms can be applied to a given black-box optimization problem without knowing much about the problem itself and without worrying about thresholds or any other parameters.

Although BOA and hBOA do not rely on problem-specific knowledge, it is fairly straightforward to incorporate prior knowledge about the problem into these algorithms to further improve their efficiency. Prior knowledge of various forms can be incorporated, including information about the problem structure or promising solutions found in previous runs.

For practitioners seeking robust, efficient, and scalable optimization techniques, BOA and hBOA represent a significant step forward. BOA and hBOA do not require the user to be an expert in the field of optimization, nor do they require the user to provide complete information about the structure of the problem and its properties. Despite that, BOA and hBOA can achieve competitive or better performance than specialized methods that fully rely on problem-specific knowledge. Furthermore, prior problem-specific knowledge can be incorporated at no extra cost, and the algorithms can be easily combined with specialized local searchers.

BOA and hBOA should have a large impact on the research in genetic and evolutionary computation and computational optimization in general. First of all, BOA provides elegant and powerful solution for linkage learning (automatic identification of building blocks) in genetic and evolutionary algorithms. Second, hBOA extends the basic approach to linkage learning to solve difficult hierarchical problems, which are practically unsolvable by other known optimization methods. Finally, BOA and hBOA replace specialized heuristics that are often incorporated into genetic and evolutionary optimization techniques to improve their performance on difficult problems by rigorous methods of statistics and probability theory.

References

1. Ackley, D. H. (1987). An empirical study of bit vector function optimization. *Genetic Algorithms and Simulated Annealing*, pages 170–204.
2. Albert, L. A. (2001). Efficient genetic algorithms using discretization scheduling. Master's thesis, University of Illinois at Urbana-Champaign, Department of General Engineering, Urbana, IL.
3. Anglès d'Auriac, J. C., Preissmann, M., and Rammal, R. (1985). The random field Ising model: Algorithmic complexity and phase transition. *J. de Physique Lett.*, 46:L173–L180.
4. Asoh, H. and Mühlenbein, H. (1994). On the mean convergence time of evolutionary algorithms without selection and mutation. *Parallel Problem Solving from Nature*, pages 88–97.
5. Bäck, T. (1995). Generalized convergence models for tournament – and (μ, λ) – selection. *Proceedings of the International Conference on Genetic Algorithms (ICGA-95)*, pages 2–8.
6. Baker, J. E. (1985). Adaptive selection methods for genetic algorithms. *Proceedings of the International Conference on Genetic Algorithms (ICGA-85)*, pages 101–111.
7. Baluja, S. (1994). Population-based incremental learning: A method for integrating genetic search based function optimization and competitive learning. Tech. Rep. No. CMU-CS-94-163, Carnegie Mellon University, Pittsburgh, PA.
8. Baluja, S. and Davies, S. (1997). Using optimal dependency-trees for combinatorial optimization: Learning the structure of the search space. *Proceedings of the International Conference on Machine Learning*, pages 30–38.
9. Baluja, S. and Davies, S. (1998). Fast probabilistic modeling for combinatorial optimization. *Proceedings of the Fifteenth National Conference on Artificial Intelligence (AAAI-98)*, pages 469–476.
10. Barahona, F. (1982). On the computational complexity of Ising spin glass models. *Journal of Physics A: Mathematical, Nuclear and General*, 15(10):3241–3253.
11. Booker, L. B. (1982). *Intelligent behavior as an adaptation to the task environment*. PhD thesis, The University of Michigan. (University Microfilms No. 8214966).

12. Bosman, P. and Thierens, D. (2001). Exploiting gradient information in continuous iterated density estimation evolutionary algorithms. *Proceedings of the Belgium-Netherlands Conference on Artificial Intelligence (BNAIC-2001)*, pages 69–76.

13. Bosman, P. A. and Thierens, D. (2000a). Continuous iterated density estimation evolutionary algorithms within the IDEA framework. *Workshop Proceedings of the Genetic and Evolutionary Computation Conference (GECCO-2000)*, pages 197–200.

14. Bosman, P. A. and Thierens, D. (2000b). Mixed IDEAs. Utrecht University Technical Report UU-CS-2000-45, Utrecht University, Utrecht, Netherlands.

15. Bosman, P. A. N. and Thierens, D. (1999). Linkage information processing in distribution estimation algorithms. *Proceedings of the Genetic and Evolutionary Computation Conference (GECCO-99)*, I:60–67.

16. Bulmer, M. G. (1985). *The mathematical theory of quantitative genetics*. Oxford University Press, Oxford.

17. Cantú-Paz, E. (2001). Supervised and unsupervised discretization methods for evolutionary algorithms. *Workshop Proceedings of the Genetic and Evolutionary Computation Conference (GECCO-2001)*, pages 213–216.

18. Cantú-Paz, E. (2002). Order statistics and selection methods of evolutionary algorithms. *Information Processing Letters*, 82(1):15–22.

19. Cavicchio, Jr., D. J. (1970). *Adaptive search using simulated evolution*. Unpublished doctoral dissertation, University of Michigan, Ann Arbor, MI. (University Microfilms No. 25-0199).

20. Ceroni, A., Pelikan, M., and Goldberg, D. E. (2001). Convergence-time models for the simple genetic algorithm with finite population. IlliGAL Report No. 2001028, University of Illinois at Urbana-Champaign, Illinois Genetic Algorithms Laboratory, Urbana, IL.

21. Cheeseman, P., Kanefsky, B., and Taylor, W. M. (1991). Where the really hard problems are. *Proceedings of the International Joint Conference on Artificial Intelligence (IJCAI-91)*, pages 331–337.

22. Chickering, D. M., Geiger, D., and Heckerman, D. (1994). Learning Bayesian networks is NP-hard. Technical Report MSR-TR-94-17, Microsoft Research, Redmond, WA.

23. Chickering, D. M., Heckerman, D., and Meek, C. (1997). A Bayesian approach to learning Bayesian networks with local structure. Technical Report MSR-TR-97-07, Microsoft Research, Redmond, WA.

24. Chow, C. and Liu, C. (1968). Approximating discrete probability distributions with dependence trees. *IEEE Transactions on Information Theory*, 14:462–467.

25. Cohoon, J. P., Hegde, S. U., Martin, W. N., and Richards, D. (1987). Punctuated equilibria: A parallel genetic algorithm. *Proceedings of the International Conference on Genetic Algorithms (ICGA-87)*, pages 148–154.

26. Collins, R. J. and Jefferson, D. R. (1991). Selection in massively parallel genetic algorithms. *Proceedings of the International Conference on Genetic Algorithms (ICGA-91)*, pages 249–256.

27. Cooper, G. F. and Herskovits, E. H. (1992). A Bayesian method for the induction of probabilistic networks from data. *Machine Learning*, 9:309–347.

28. Culberson, J. C. (1992). Genetic invariance: A new paradigm for genetic algorithm design. Unpublished manuscript.

29. Davidor, Y. (1991). A naturally occuring niche and species phenomenon: The model and first results. *Proceedings of the International Conference on Genetic Algorithms (ICGA-91)*, pages 257–263.

30. Davies, S. and Moore, A. (1999). Using Bayesian networks for lossless compression in data mining. *Proceedings of the International Conference on Knowledge Discovery & Data Mining (KDD-99)*, pages 387–391.

31. Davis, M., Logemann, G., and Loveland, D. (1962). A machine program for theorem proving. *Communications of the ACM*, 5(7):394–397.

32. Dayal, P., Trebst, S., Wessel, S., Würtz, D., Troyer, M., Sabhapandit, S., and Coppersmith, S. N. (2004). Performance limitations of flat histogram methods. *Physical Review Letters*. In press.

33. De Bonet, J. S., Isbell, C. L., and Viola, P. (1997). MIMIC: Finding optima by estimating probability densities. *Advances in neural information processing systems (NIPS-97)*, 9:424–431.

34. De Jong, K. A. (1975). *An analysis of the behavior of a class of genetic adaptive systems*. PhD thesis, University of Michigan, Ann Arbor. (University Microfilms No. 76-9381).

35. De Simone, C., Diehl, M., Jünger, M., and Reinelt, G. (1996). Exact ground states of two-dimensional $+ - J$ Ising spin glasses. *Journal of Statistical Physics*, 84:1363–1371.

36. Deb, K. (2001). *Multi-objective optimization using evolutionary algorithms*. John Wiley & Sons, Chichester, UK.

37. Deb, K. and Goldberg, D. E. (1989). An investigation of niche and species formation in genetic function optimization. *Proceedings of the International Conference on Genetic Algorithms (ICGA-89)*, pages 42–50.

38. Deb, K. and Goldberg, D. E. (1991). Analyzing deception in trap functions. IlliGAL Report No. 91009, University of Illinois at Urbana-Champaign, Illinois Genetic Algorithms Laboratory, Urbana, IL.

39. Edmonds, J. (1967). Optimum branching. *J. Res. Nat. Bur. Standards*, 71B:233–240.

40. Eldredge, N. and Gould, S. J. (1972). Punctuated equilibria: An alternative to phyletic gradualism. In Schopf, T., editor, *Paleobiology*, pages 82–115. Freeman & Company, San Francisco, CA.

41. Etxeberria, R. and Larrañaga, P. (1999). Global optimization using Bayesian networks. In Rodriguez, A. A. O., Ortiz, M. R. S., and Hermida, R. S., editors, *Second Symposium on Artificial Intelligence (CIMAF-99)*, pages 332–339, Habana, Cuba. Institute of Cybernetics, Mathematics, and Physics and Ministry of Science, Technology and Environment.

42. Feller, W. (1970). *An introduction to probability theory and its applications*. Wiley, New York, NY.

43. Fonseca, C. M. and Fleming, P. J. (1993). Genetic algorithms for multiobjective optimization: Formulation, discussion and generalization. *Proceedings of the International Conference on Genetic Algorithms (ICGA-93)*, pages 416–423.

44. Friedman, N. and Goldszmidt, M. (1999). Learning Bayesian networks with local structure. In Jordan, M. I., editor, *Graphical models*, pages 421–459. MIT Press, Cambridge, MA.

45. Friedman, N. and Yakhini, Z. (1996). On the sample complexity of learning Bayesian networks. *Proceedings of the Conference on Uncertainty in Artificial Intelligence (UAI-96)*, pages 274–282.

46. Gallagher, M., Frean, M., and Downs, T. (1999). Real-valued evolutionary optimization using a flexible probability density estimator. *Proceedings of the Genetic and Evolutionary Computation Conference (GECCO-99)*, 1:840–846.

47. Galluccio, A. and Loebl, M. (1999a). A theory of Pfaffian orientations. I. Perfect matchings and permanents. *Electronic Journal of Combinatorics*, 6(1). Research Paper 6.

48. Galluccio, A. and Loebl, M. (1999b). A theory of Pfaffian orientations. II. T-joins, k-cuts, and duality of enumeration. *Electronic Journal of Combinatorics*, 6(1). Research Paper 7.

49. Geiger, D. and Heckerman, D. (1996). Beyond Bayesian networks: Similarity networks and Bayesian multinets. *Artificial Intelligence*, 82:45–74.

50. Geiger, D., Heckerman, D., and Meek, C. (1996). Asymptotic model selection for directed networks with hidden variables. *Proceedings of the Conference on Uncertainty in Artificial Intelligence (UAI-96)*, pages 158–168.

51. Gent, I., Hoos, H. H., Prosser, P., and Walsh, T. (1999). Morphing: Combining structure and randomness. *Proceedings of the American Association of Artificial Intelligence (AAAI-99)*, pages 654–660.

52. Goldberg, D. E. (1983). Computer-aided gas pipeline operation using genetic algorithms and rule learning. *Dissertation Abstracts International*, 44(10):3174B. Doctoral dissertation, University of Michigan.

53. Goldberg, D. E. (1989a). *Genetic algorithms in search, optimization, and machine learning*. Addison-Wesley, Reading, MA.

54. Goldberg, D. E. (1989b). Sizing populations for serial and parallel genetic algorithms. *Proceedings of the International Conference on Genetic Algorithms (ICGA-89)*, pages 70–79.

55. Goldberg, D. E. (1998). Four keys to understanding building-block difficulty. Presented in Projet FRACTALES Seminar at I.N.R.I.A. Rocquencourt, Le Chesnay, Cedex.

56. Goldberg, D. E. (1999). Genetic and evolutionary algorithms in the real world. IlliGAL Report No. 99013, University of Illinois at Urbana-Champaign, Illinois Genetic Algorithms Laboratory, Urbana, IL.

57. Goldberg, D. E. (2002). *The design of innovation: Lessons from and for competent genetic algorithms*, volume 7 of *Genetic Algorithms and Evolutionary Computation*. Kluwer Academic Publishers.

58. Goldberg, D. E., Deb, K., and Clark, J. H. (1992). Genetic algorithms, noise, and the sizing of populations. *Complex Systems*, 6:333–362.

59. Goldberg, D. E. and Richardson, J. (1987). Genetic algorithms with sharing for multimodal function optimization. *Proceedings of the International Conference on Genetic Algorithms (ICGA-87)*, pages 41–49.

60. Goldberg, D. E. and Rudnick, M. (1991). Genetic algorithms and the variance of fitness. *Complex Systems*, 5(3):265–278.

61. Goldberg, D. E., Sastry, K., and Latoza, T. (2001). On the supply of building blocks. *Proceedings of the Genetic and Evolutionary Computation Conference (GECCO-2001)*, pages 336–342.

62. Goldberg, D. E. and Segrest, P. (1987). Finite Markov chain analysis of genetic algorithms. *Proceedings of the International Conference on Genetic Algorithms (ICGA-87)*, pages 1–8.

63. Gonzalez, C., Lozano, J., and Larrañaga, P. (2001). Analyzing the PBIL algorithm by means of discrete dynamical systems. *Complex Systems*, 4(12):465–479.

64. Gorges-Schleuter, M. (1989). ASPARAGOS: An asynchronous parallel genetic optimization strategy. *Proceedings of the International Conference on Genetic Algorithms (ICGA-89)*, pages 422–428.

65. Gottlieb, J., Marchiori, E., and Rossi, C. (2002). Evolutionary algorithms for the satisfiability problem. *Evolutionary Computation*, 10(1):35–50.

66. Grosso, P. B. (1985). *Computer simulations of genetic adaptation: Parallel subcomponent interaction in a multilocus model.* Unpublished doctoral dissertation, University of Michigan. (University Microfilms No. 8520908).

67. Grünwald, P. (1998). *The minimum description length principle and reasoning under uncertainty.* PhD thesis, University of Amsterdam, Amsterdam, Netherlands.

68. Handley, S. (1994). On the use of a directed acyclic graph to represent a population of computer programs. *Proceedings of the International Conference on Evolutionary Computation (ICEC-94)*, pages 154–159.

69. Hansen, N., Ostermeier, A., and Gawelczyk, A. (1995). On the adaptation of arbitrary normal mutation distributions in evolution strategies: The generating set adaptation. *Proceedings of the International Conference on Genetic Algorithms (ICGA-95)*, pages 57–64.

70. Harik, G. (1999). Linkage learning via probabilistic modeling in the ECGA. IlliGAL Report No. 99010, University of Illinois at Urbana-Champaign, Illinois Genetic Algorithms Laboratory, Urbana, IL.

71. Harik, G., Cantú-Paz, E., Goldberg, D. E., and Miller, B. L. (1999a). The gambler's ruin problem, genetic algorithms, and the sizing of populations. *Evolutionary Computation*, 7(3):231–253.

72. Harik, G. and Lobo, F. G. (1999). A parameter-less genetic algorithm. *Proceedings of the Genetic and Evolutionary Computation Conference (GECCO-99)*, I:258–265.

73. Harik, G. R. (1994). Finding multiple solutions in problems of bounded difficulty. IlliGAL Report No. 94002, University of Illinois at Urbana-Champaign, Urbana, IL.

74. Harik, G. R., Cantú-Paz, E., Goldberg, D. E., and Miller, B. L. (1997). The gambler's ruin problem, genetic algorithms, and the sizing of populations. *Proceedings of the International Conference on Evolutionary Computation (ICEC-97)*, pages 7–12.

75. Harik, G. R., Lobo, F. G., and Goldberg, D. E. (1998). The compact genetic algorithm. *Proceedings of the International Conference on Evolutionary Computation (ICEC-98)*, pages 523–528.

76. Harik, G. R., Lobo, F. G., and Goldberg, D. E. (1999b). The compact genetic algorithm. *IEEE Transactions on Evolutionary Computation*, 3(4):287–297.

77. Hartmann, A. K. (1996). Cluster-exact approximation of spin glass ground states. *Physica A*, 224(3–4):480–488.

78. Hartmann, A. K. (2001). Ground-state clusters of two, three and four-dimensional +/-J Ising spin glasses. *Phys. Rev. E*, 63:016106.

79. Heckerman, D., Geiger, D., and Chickering, D. M. (1994). Learning Bayesian networks: The combination of knowledge and statistical data. Technical Report MSR-TR-94-09, Microsoft Research, Redmond, WA.

80. Henrion, M. (1988). Propagation of uncertainty in Bayesian networks by logic sampling. In Lemmer, J. F. and Kanal, L. N., editors, *Uncertainty in Artificial Intelligence*, pages 149–163. Elsevier, Amsterdam, London, New York.

81. Höhfeld, M. and Rudolph, G. (1997). Towards a theory of population-based incremental learning. *Proceedings of the International Conference on Evolutionary Computation (ICEC-97)*, pages 1–6.

82. Holland, J. (1973). Genetic algorithms and the optimal allocation of trials. *SIAM Journal of Computing*, 2(2):88–105.

83. Holland, J. H. (1975). *Adaptation in natural and artificial systems*. University of Michigan Press, Ann Arbor, MI.

84. Holland, J. H. (2000). Building blocks, cohort genetic algorithms, and hyperplane-defined functions. *Evolutionary Computation*, 8(4):373–391.

85. Hollstein, R. B. (1971). *Artificial genetic adaptation in computer control systems*. PhD thesis, University of Michigan. (University Microfilms No. 71-23,773).

86. Horn, J. (1993). Finite Markov chain analysis of genetic algorithms with niching. *Proceedings of the International Conference on Genetic Algorithms (ICGA-93)*, pages 110–117.

87. Horn, J. and Nafpliotis, N. (1993). Multiobjective optimization using the niched pareto genetic algorithm. IlliGAL Report No. 93005, University of Illinois at Urbana-Champaign, Illinois Genetic Algorithms Laboratory, Urbana, IL.

88. Howard, R. A. and Matheson, J. E. (1981). Influence diagrams. In Howard, R. A. and Matheson, J. E., editors, *Readings on the principles and applications of decision analysis*, volume II, pages 721–762. Strategic Decisions Group, Menlo Park, CA.

89. Kardar, M. and Saul, L. (1994). The 2D $+/-$J Ising spin glass: Exact partition functions in polynomial time. *Nucl. Phys. B*, 432:641–667.

90. Khan, N., Goldberg, D. E., and Pelikan, M. (2002). Multiobjective Bayesian optimization algorithm. IlliGAL Report No. 2002009, University of Illinois at Urbana-Champaign, Illinois Genetic Algorithms Laboratory, Urbana, IL.

91. Kirkpatrick, S., Gelatt, C. D., and Vecchi, M. P. (1983). Optimization by simulated annealing. *Science*, 220:671–680.

92. Koza, J. R. (1992). *Genetic programming: On the programming of computers by means of natural selection*. MIT Press, Cambridge, MA.

93. Koza, J. R. (1994). *Genetic programming II: Automatic discovery of reusable programs*. MIT Press, Cambridge, MA.

94. Kullback, S. and Leibler, R. A. (1951). On information and sufficiency. *Annals of Math. Stats.*, 22:79–86.

95. Kvasnicka, V., Pelikan, M., and Pospichal, J. (1996). Hill climbing with learning (An abstraction of genetic algorithm). *Neural Network World*, 6:773–796.

96. Larrañaga, P., Etxeberria, R., Lozano, J., and Pena, J. (2000a). Combinatorial optimization by learning and simulation of Bayesian networks. *Proceedings of the Uncertainty in Artificial Intelligence (UAI-2000)*, pages 343–352.

97. Larrañaga, P., Etxeberria, R., Lozano, J. A., and Pena, J. M. (2000b). Optimization in continuous domains by learning and simulation of Gaussian networks. *Workshop Proceedings of the Genetic and Evolutionary Computation Conference (GECCO-2000)*, pages 201–204.

98. Laumanns, M. and Ocenasek, J. (2002). Bayesian optimization algorithms for multi-objective optimization. *Parallel Problem Solving from Nature*, pages 298–307.

99. Li, C. M. and Anbulagan (1997). Heuristics based on unit propagation for satisfiability problems. *Proceedings of the International Joint Conference on Artificial Intelligence (IJCAI-97)*, pages 366–371.

100. Li, J. and Aickelin, U. (2003). A Bayesian optimization algorithm for the nurse scheduling problem. *Proceedings of the Congress on Evolutionary Computation (CEC-2003)*, pages 2149–2156.

101. Lobo, F. G., Goldberg, D. E., and Pelikan, M. (2000). Time complexity of genetic algorithms on exponentially scaled problems. *Proceedings of the Genetic and Evolutionary Computation Conference (GECCO-2000)*, pages 151–158.

102. Looks, M., Goertzel, B., and Pennachin, C. (2004). Learning computer programs with the Bayesian optimization algorithm. Personal communication.

103. Mahfoud, S. W. (1992). Crowding and preselection revisited. *Parallel Problem Solving from Nature*, pages 27–36.

104. Marascuilo, L. A. and McSweeney, M. (1977). *Nonparametric and distribution-free methods for the social sciences*. Brooks/Cole Publishing Company, CA.

105. Mauldin, M. L. (1984). Maintaining diversity in genetic search. In Brachman, R. J., editor, *Proceedings of the National Conference on Artificial Intelligence*, pages 247–250, Austin, TX. William Kaufmann.

106. Mengshoel, O. J. and Goldberg, D. E. (1999). Probabilistic crowding: Deterministic crowding with probabilisitic replacement. *Proceedings of the Genetic and Evolutionary Computation Conference (GECCO-99)*, I:409–416.

107. Miller, B. L. and Goldberg, D. E. (1996). Genetic algorithms, selection schemes, and the varying effects of noise. *Evolutionary Computation*, 4(2):113–131.

108. Mitchell, M. (1996). *An introduction to genetic algorithms*. MIT Press, Cambridge, MA.

109. Mitchell, M., Forrest, S., and Holland, J. H. (1992). The royal road for genetic algorithms: Fitness landscapes and GA performance. *Toward a Practice of Autonomous Systems: Proceedings of the First European Conference on Artificial Life*, pages 245–254.

110. Monien, B. and Sudborough, I. H. (1988). Min cut is NP-complete for edge weighted trees. *Theoretical Computer Science*, 58(1–3):209–229.

111. Mühlenbein, H. (1991). Evolution in time and space-The parallel genetic algorithm. *Foundations of Genetic Algorithms*, pages 316–337.

112. Mühlenbein, H. (1992). How genetic algorithms really work: I. Mutation and hillclimbing. *Parallel Problem Solving from Nature*, pages 15–25.

113. Mühlenbein, H. (1997). The equation for response to selection and its use for prediction. *Evolutionary Computation*, 5(3):303–346.

114. Mühlenbein, H. and Mahnig, T. (1998). Convergence theory and applications of the factorized distribution algorithm. *Journal of Computing and Information Technology*, 7(1):19–32.

115. Mühlenbein, H. and Mahnig, T. (1999). FDA – A scalable evolutionary algorithm for the optimization of additively decomposed functions. *Evolutionary Computation*, 7(4):353–376.

116. Mühlenbein, H., Mahnig, T., and Rodriguez, A. O. (1999). Schemata, distributions and graphical models in evolutionary optimization. *Journal of Heuristics*, 5:215–247.

117. Mühlenbein, H. and Paaß, G. (1996). From recombination of genes to the estimation of distributions I. Binary parameters. *Parallel Problem Solving from Nature*, pages 178–187.

118. Mühlenbein, H. and Schlierkamp-Voosen, D. (1993). Predictive models for the breeder genetic algorithm: I. Continuous parameter optimization. *Evolutionary Computation*, 1(1):25–49.

119. Naudts, B. and Naudts, J. (1998). The effect of spin-flip symmetry on the performance of the simple GA. *Parallel Problem Solving from Nature*, pages 67–76.

120. Ocenasek, J. and Pelikan, M. (2004). Parallel mixed Bayesian optimization algorithm: A scaleup analysis. Workshop Proceedings of the Genetic and Evolutionary Computation Conference (GECCO-2004).

121. Ocenasek, J. and Schwarz, J. (2002). Estimation of distribution algorithm for mixed continuous-discrete optimization problems. In *2nd Euro-International Symposium on Computational Intelligence*, pages 227–232, Kosice, Slovakia. IOS Press.

122. Ogielsky, A. T. (1986). Integer optimization and zero-temperature fixed point in Ising random-field systems. *Phys. Rev. Lett.*, 57:1251–1254.

123. Pearl, J. (1988). *Probabilistic reasoning in intelligent systems: Networks of plausible inference*. Morgan Kaufmann, San Mateo, CA.

124. Pelikan, M. (2002). *Bayesian optimization algorithm: From single level to hierarchy*. PhD thesis, University of Illinois at Urbana-Champaign, Urbana, IL.

125. Pelikan, M. and Goldberg, D. E. (2000a). Genetic algorithms, clustering, and the breaking of symmetry. *Parallel Problem Solving from Nature*, pages 385–394.

126. Pelikan, M. and Goldberg, D. E. (2000b). Hierarchical problem solving and the Bayesian optimization algorithm. *Proceedings of the Genetic and Evolutionary Computation Conference (GECCO-2000)*, pages 275–282.

127. Pelikan, M. and Goldberg, D. E. (2001). Escaping hierarchical traps with competent genetic algorithms. *Proceedings of the Genetic and Evolutionary Computation Conference (GECCO-2001)*, pages 511–518.

128. Pelikan, M. and Goldberg, D. E. (2003). Hierarchical BOA solves Ising spin glasses and MAXSAT. *Proceedings of the Genetic and Evolutionary Computation Conference (GECCO-2003)*, II:1275–1286.

129. Pelikan, M., Goldberg, D. E., and Cantú-Paz, E. (1998). Linkage problem, distribution estimation, and Bayesian networks. IlliGAL Report No. 98013, University of Illinois at Urbana-Champaign, Illinois Genetic Algorithms Laboratory, Urbana, IL.

130. Pelikan, M., Goldberg, D. E., and Cantú-Paz, E. (1999). BOA: The Bayesian optimization algorithm. *Proceedings of the Genetic and Evolutionary Computation Conference (GECCO-99)*, I:525–532.

131. Pelikan, M., Goldberg, D. E., and Cantú-Paz, E. (2000a). Bayesian optimization algorithm, population sizing, and time to convergence. *Proceedings of the Genetic and Evolutionary Computation Conference (GECCO-2000)*, pages 275–282.

132. Pelikan, M., Goldberg, D. E., and Cantú-Paz, E. (2000b). Linkage problem, distribution estimation, and Bayesian networks. *Evolutionary Computation*, 8(3):311–341.

133. Pelikan, M., Goldberg, D. E., and Lobo, F. G. (2002a). A survey of optimization by building and using probabilistic models. *Computational Optimization and Applications*, 21(1):5–20.

134. Pelikan, M., Goldberg, D. E., and Sastry, K. (2001a). Bayesian optimization algorithm, decision graphs, and Occam's razor. *Proceedings of the Genetic and Evolutionary Computation Conference (GECCO-2001)*, pages 519–526.

135. Pelikan, M., Goldberg, D. E., and Tsutsui, S. (2002b). Combining the strengths of the Bayesian optimization algorithm and adaptive evolution strategies. *Proceedings of the Genetic and Evolutionary Computation Conference (GECCO-2002)*, pages 512–519.

136. Pelikan, M., Goldberg, D. E., and Tsutsui, S. (2003). Getting the best of both worlds: Discrete and continuous genetic and evolutionary algorithms in concert. *Information Sciences*, 156(3–4):36–45.

137. Pelikan, M. and Lin, T.-K. (2004). The parameter-less hierarchical BOA. *Proceedings of the Genetic and Evolutionary Computation Conference (GECCO-2004)*, pages 24–35.

138. Pelikan, M. and Lobo, F. G. (1999). Parameter-less genetic algorithm: A worst-case time and space complexity analysis. IlliGAL Report No. 99014, University of Illinois at Urbana-Champaign, Illinois Genetic Algorithms Laboratory, Urbana, IL.

139. Pelikan, M. and Mühlenbein, H. (1999). The bivariate marginal distribution algorithm. *Advances in Soft Computing – Engineering Design and Manufacturing*, pages 521–535.

140. Pelikan, M., Sastry, K., and Goldberg, D. E. (2001b). Evolutionary algorithms + graphical models = scalable black-box optimization. IlliGAL Report No. 2001029, Illinois Genetic Algorithms Laboratory, University of Illinois at Urbana-Champaign, Urbana, IL.

141. Pelikan, M., Sastry, K., and Goldberg, D. E. (2002c). Scalability of the Bayesian optimization algorithm. *International Journal of Approximate Reasoning*, 31(3):221–258.

142. Perry, Z. A. (1984). Experimental study of speciation in ecological niche theory using genetic algorithms. *Dissertation Abstracts International*, 45(12):3870B. (University Microfilms No. 8502912).

143. Poli, R., Langdon, W., and O'Reilly, U.-M. (1998). Analysis of schema variance and short term extinction likelihoods. *Proceedings of the Genetic Programming Conference (GP-98)*, pages 284–292.

144. Prim, R. (1957). Shortest connection networks and some generalizations. *Bell Systems Technical Journal*, 36:1389–1401.

145. Rana, S. and Whitley, D. L. (1998). Genetic algorithm behavior in the MAXSAT domain. *Parallel Problem Solving from Nature*, pages 785–794.

146. Rechenberg, I. (1973). *Evolutionsstrategie: Optimierung technischer Systeme nach Prinzipien der biologischen Evolution*. Frommann-Holzboog, Stuttgart.

147. Rechenberg, I. (1994). *Evolutionsstrategie '94*. Frommann-Holzboog Verlag, Stuttgart.

148. Reeves, C. (1993). Using genetic algorithms with small populations. *Proceedings of the International Conference on Genetic Algorithms (ICGA-93)*, pages 92–99.

149. Rissanen, J. J. (1978). Modelling by shortest data description. *Automatica*, 14:465–471.

150. Rissanen, J. J. (1989). *Stochastic complexity in statistical inquiry.* World Scientific Publishing Co, Singapore.
151. Rissanen, J. J. (1996). Fisher information and stochastic complexity. *IEEE Transactions on Information Theory*, 42(1):40–47.
152. Rothlauf, F. (2001). *Towards a theory of representations for genetic and evolutionary algorithms: Development of basic concepts and their application to binary and tree representations.* PhD thesis, University of Bayreuth, Beyreuth, Germany.
153. Rothlauf, F. (2002). *Representations for genetic and evolutionary algorithms.* Studies in Fuzziness and Soft Computing. Springer-Verlag, Heidelberg.
154. Rothlauf, F., Goldberg, D. E., and Heinzl, A. (2000). Bad codings and the utility of well-designed genetic algorithms. IlliGAL Report No. 200007, University of Illinois at Urbana-Champaign, Illinois Genetic Algorithms Laboratory, Urbana, IL.
155. Rudlof, S. and Köppen, M. (1996). Stochastic hill climbing with learning by vectors of normal distributions. In *First On-line Workshop on Soft Computing*, Nagoya, Japan.
156. Rudnick, M. W. (1992). *Genetic algorithms and fitness variance with application to automated design of artificial neural networks.* PhD thesis, Oregon Graduate Institute of Science & Technology, Beaverton, OR.
157. Salustowicz, R. and Schmidhuber, J. (1998). H-PIPE: Facilitating hierarchical program evolution through skip nodes. Technical Report IDSIA-08-98, Instituto Dalle Molle di Studi sull' Intelligenza Artificiale (IDSIA), Lugano, Switzerland.
158. Salustowicz, R. P. and Schmidhuber, J. (1997a). Probabilistic incremental program evolution. *Evolutionary Computation*, 5(2):123–141.
159. Salustowicz, R. P. and Schmidhuber, J. (1997b). Probabilistic incremental program evolution: Stochastic search through program space. *Proceedings of the European Conference of Machine Learning (ECML-97)*, 1224:213–220.
160. Sastry, K. (2001). Efficient atomic cluster optimization using a hybrid extended compact genetic algorithm with seeded population. IlliGAL Report No. 2001018, University of Illinois at Urbana-Champaign, Illinois Genetic Algorithms Laboratory, Urbana, IL.
161. Sastry, K. and Goldberg, D. E. (2000). On extended compact genetic algorithm. IlliGAL Report No. 2000026, University of Illinois at Urbana-Champaign, Illinois Genetic Algorithms Laboratory, Urbana, IL.
162. Sastry, K. and Goldberg, D. E. (2003). Probabilistic model building and competent genetic programming. In Riolo, R. L. and Worzel, B., editors, *Genetic Programming Theory and Practise*, chapter 13, pages 205–220. Kluwer.
163. Schaffer, J. D. (1984). *Some experiments in machine learning using vector evaluated genetic algorithms.* PhD thesis, Vanderbilt University, Nashville, Tennessee. (University Microfilms No. 85-22492).
164. Schwarz, G. (1978). Estimating the dimension of a model. *The Annals of Statistics*, 6:461–464.
165. Schwarz, J. and Ocenasek, J. (1999). Experimental study: Hypergraph partitioning based on the simple and advanced algorithms BMDA and BOA. *Proceedings of the International Conference on Soft Computing*, pages 124–130.
166. Schwarz, J. and Ocenasek, J. (2000). A problem-knowledge based evolutionary algorithm KBOA for hypergraph partitioning. In *Proceedings of the*

Fourth Joint Conference on Knowledge-Based Software Engineering, pages 51–58, Brno, Czech Republic. IO Press.

167. Schwefel, H.-P. (1977). *Numerische Optimierung von Computer–Modellen mittels der Evolutionsstrategie*, volume 26 of *Interdisciplinary Systems Research*. Birkhäuser, Basle, Switzerland.

168. Sebag, M. and Ducoulombier, A. (1998). Extending population-based incremental learning to continuous search spaces. *Parallel Problem Solving from Nature*, pages 418–427.

169. Selman, B., Levesque, H. J., and Mitchell, D. (1992). A new method for solving hard satisfiability problems. *Proceedings of the National Conference on Artificial Intelligence (AAAI-92)*, pages 440–446.

170. Servet, I., Trave-Massuyes, L., and Stern, D. (1997). Telephone network traffic overloading diagnosis and evolutionary computation techniques. *Proceedings of the European Conference on Artificial Evolution (AE-97)*, pages 137–144.

171. Simon, H. A. (1968). *The sciences of the artificial*. The MIT Press, Cambridge, MA.

172. Thierens, D. (1995). *Analysis and design of genetic algorithms*. PhD thesis, Katholieke Universiteit Leuven, Leuven, Belgium.

173. Thierens, D. (1997). Selection schemes, elitist recombination, and selection intensity. *Proceedings of the International Conference on Genetic Algorithms (ICGA-97)*, pages 152–159.

174. Thierens, D. and Bosman, P. A. N. (2001). Multi-objective mixture-based iterated density estimation evolutionary algorithms. *Proceedings of the Genetic and Evolutionary Computation Conference (GECCO-2001)*, pages 663–670.

175. Thierens, D. and Goldberg, D. (1994). Convergence models of genetic algorithm selection schemes. *Parallel Problem Solving from Nature*, pages 116–121.

176. Thierens, D. and Goldberg, D. E. (1993). Mixing in genetic algorithms. *Proceedings of the International Conference on Genetic Algorithms (ICGA-93)*, pages 38–45.

177. Thierens, D., Goldberg, D. E., and Pereira, A. G. (1998). Domino convergence, drift, and the temporal-salience structure of problems. *Proceedings of the International Conference on Evolutionary Computation (ICEC-98)*, pages 535–540.

178. Tsutsui, S., Pelikan, M., and Goldberg, D. E. (2001). Evolutionary algorithm using marginal histogram models in continuous domain. *Workshop Proceedings of the Genetic and Evolutionary Computation Conference (GECCO-2001)*, pages 230–233.

179. Van Hoyweghen, C. (2001). Detecting spin-flip symmetry in optimization problems. In Kallel, L., Naudts, B., and Rogers, A., editors, *Theoretical Aspects of Evolutionary Computing*, pages 423–437. Springer, Berlin.

180. Watson, R. A. (2001). Analysis of recombinative algorithms on a non-separable building-block problem. *Foundations of Genetic Algorithms*, pages 69–90.

181. Watson, R. A., Hornby, G. S., and Pollack, J. B. (1998). Modeling building-block interdependency. *Parallel Problem Solving from Nature*, pages 97–106.

182. Wright, S. (1968). *Evolution and the genetics of populations: A treatise*. University of Chicago Press.

183. Zhang, B.-T. and Shin, S.-Y. (2000). Bayesian evolutionary optimization using Helmholtz machines. *Parallel Problem Solving from Nature*, pages 827–836.

Index